Edexcel AS Physics

Miles Hudson

STUDENTS' BOOK

How to use this book

This book contains a number of great features that will help you find your way around your AS Physics course and support your learning.

Introductory pages

Each topic has two introductory pages to help you identify how the main text is arranged to cover all that you need to learn. The left-hand page gives a brief summary of the topic, linking the content to three key areas of How Science Works:

What are the theories? What is the evidence? What are the implications?

The right-hand page of the introduction consists of a topic map that shows you how all the required content of the Edexcel specification for that topic is covered in the chapters, and how that content all interlinks. Links to other topics are also shown, including where previous knowledge is built on within the topic.

Main text

The main part of the book covers all you need to learn for your course. The text is supported by many diagrams and photographs that will help you understand the concepts you need to learn.

Key terms in the text are shown in bold type. These terms are defined in the interactive glossary that can be found on the software using the 'search glossary' feature.

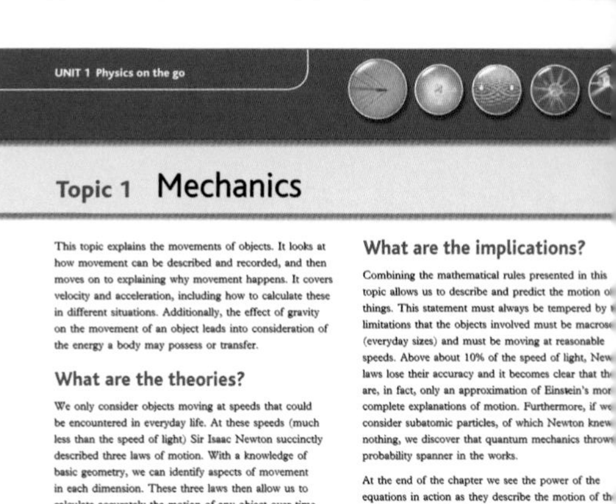
UNIT 1 Physics on the go

Topic 1 Mechanics

This topic explains the movements of objects. It looks at how movement can be described and recorded, and then moves on to explaining why movement happens. It covers velocity and acceleration, including how to calculate these in different situations. Additionally, the effect of gravity on the movement of an object leads into consideration of the energy a body may possess or transfer.

What are the theories?

We only consider objects moving at speeds that could be encountered in everyday life. At these speeds (much less than the speed of light) Sir Isaac Newton succinctly described three laws of motion. With a knowledge of basic geometry, we can identify aspects of movement in each dimension. These three laws then allow us to calculate accurately the motion of any object over time and in three dimensions.

There are also equations for calculating kinetic energy and gravitational potential energy, and the transfer of energy when a force is used to cause the transfer. These formulae and Newton's laws can be used together to work out everything we might wish to know about the movement of any everyday object in any everyday situation.

What is the evidence?

Newton's laws of motion have been constantly under test by scientists ever since he published them in 1687. Within constraints established by Einstein in the early twentieth century, Newton's laws have always correctly described the relationships between data collected. You may have a chance to confirm Newton's laws in experiments of your own. With modern ICT recording of data, the reliability of such experiments is now much improved over traditional methods.

Whilst it is difficult for scientists to describe or identify the exact nature of energy, the equations that describe energy relationships have also consistently held up to experimental scrutiny.

What are the implications?

Combining the mathematical rules presented in this topic allows us to describe and predict the motion of things. This statement must always be tempered by the limitations that the objects involved must be macrosc (everyday sizes) and must be moving at reasonable speeds. Above about 10% of the speed of light, New laws lose their accuracy and it becomes clear that th are, in fact, only an approximation of Einstein's mor complete explanations of motion. Furthermore, if we consider subatomic particles, of which Newton knew nothing, we discover that quantum mechanics throw probability spanner in the works.

At the end of the chapter we see the power of the equations in action as they describe the motion of th ball in a game of hockey.

The map opposite shows you all the knowledge and you need to have by the end of this topic. The colo in each box shows which chapter they are covered in and the numbers refer to the sections in the Edexcel specification.

8

Introductory pages

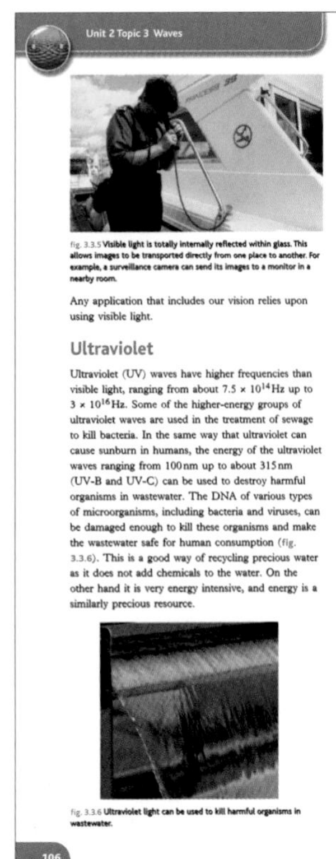
Unit 2 Topic 3 Waves

fig. 3.3.5 Visible light is totally internally reflected within glass. This allows images to be transported directly from one place to another. For example, a surveillance camera can send its images to a monitor in a nearby room.

Any application that includes our vision relies upon using visible light.

Ultraviolet

Ultraviolet (UV) waves have higher frequencies than visible light, ranging from about 7.5×10^{14} Hz up to 3×10^{16} Hz. Some of the higher-energy groups of ultraviolet waves are used in the treatment of sewage to kill bacteria. In the same way that ultraviolet can cause sunburn in humans, the energy of the ultraviolet waves ranging from 100 nm up to about 315 nm (UV-B and UV-C) can be used to destroy harmful organisms in wastewater. The DNA of various types of microorganisms, including bacteria and viruses, can be damaged enough to kill these organisms and make the wastewater safe for human consumption (fig. 3.3.6). This is a good way of recycling precious water as it does not add chemicals to the water. On the other hand it is very energy intensive, and energy is a similarly precious resource.

fig. 3.3.6 Ultraviolet light can be used to kill harmful organisms in wastewater.

X-rays

Very high energy electromagnetic waves can be produced by colliding high-speed electrons with a metal target. When these electrons decelerate rapidly, they give off energy as **X-rays**. This is how a hospital X-ray machine produces its electromagnetic waves. These waves will affect photographic film, which causes the cloudy picture you may be familiar with in medical X-rays. Recently, X-ray detectors have been developed that do not rely upon a chemical reaction. These can produce digital images of the X-rays detected without needing to use up a constant supply of photographic film. X-ray scanning can also be used in any industry in which the integrity of metal parts needs to be tested – aeroplane engines, for example. The part is not damaged by the test, but can be checked for cracks and other flaws. Dense materials, like metals, will absorb X-rays, and can produce a shadow image, just like a medical X-ray image. If the metal is cracked, more X-rays will pass through without being absorbed and this will show up on the image.

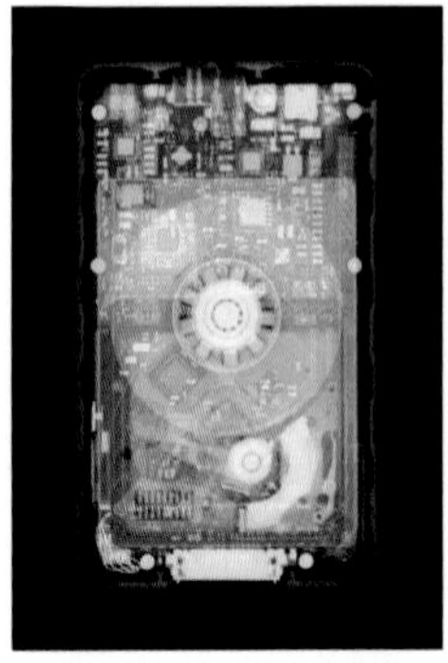
fig. 3.3.7 X-rays can show us inside metal objects without breaking them open.

106

Unit 2 Topic 3 Waves

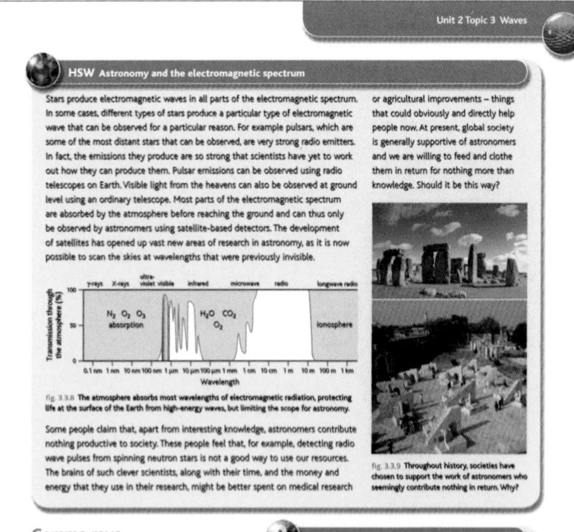
HSW Astronomy and the electromagnetic spectrum

Stars produce electromagnetic waves in all parts of the electromagnetic spectrum. In some cases, different types of stars produce a particular type of electromagnetic wave that can be observed for a particular reason. For example pulsars, which are some of the most distant stars that can be observed, are very strong radio emitters. In fact, the emissions they produce are so strong that scientists have yet to work out how they can produce them. Pulsar emissions can be observed using radio telescopes on Earth. Visible light from the heavens can also be observed at ground level using an ordinary telescope. Most parts of the electromagnetic spectrum are absorbed by the atmosphere before reaching the ground and can thus only be observed by astronomers using satellite-based detectors. The development of satellites has opened up vast new areas of research in astronomy, as it is now possible to scan the skies at wavelengths that were previously invisible.

fig. 3.3.8 The atmosphere absorbs most wavelengths of electromagnetic radiation, protecting life at the surface of the Earth from high-energy waves, but limiting the scope for astronomy.

Some people claim that, apart from interesting knowledge, astronomers contribute nothing productive to society. These people feel that, for example, detecting radio wave pulses from spinning neutron stars is not a good way to use our resources. The brains of such clever scientists, along with their time, and the money and energy that they use in their research, might be better spent on medical research or agricultural improvements – things that could obviously and directly help people now. At present, global society is generally supportive of astronomers and we are willing to feed and clothe them in return for nothing more than knowledge. Should it be this way?

fig. 3.3.9 Throughout history, societies have chosen to support the work of astronomers who seemingly contribute nothing in return. Why?

Gamma rays

Generated by energy shifts inside the nuclei of atoms, **gamma rays** form the very highest energy EM waves. At lower energies, there is a crossover between the frequency ranges of X-rays and gamma rays, but the distinction comes from how the electromagnetic waves were produced. Most gamma-ray applications come from their property of being fatal to biological cells. They are used in the sterilisation of surgical instruments and soft fruits. Gamma rays are also used to kill cancerous cells in the body. This can be a dangerous procedure, as the interaction between healthy cells and gamma rays is not very good for the body.

Questions

1 a Calculate the wavelength range of radio waves that suffer reflection by the ionosphere.
  b Explain why communications with satellites must use radio/microwaves with a wavelength shorter than 1 metre.
2 Explain why the human eye did not develop so that it could detect UV light with a wavelength of 100 nm.
3 Give a similarity and a difference between X-rays and gamma rays.
4 Draw a diagram to illustrate how X-rays could be used in a machine to produce images that could be scrutinised to detect flaws in vehicle engine parts.

107

Main text

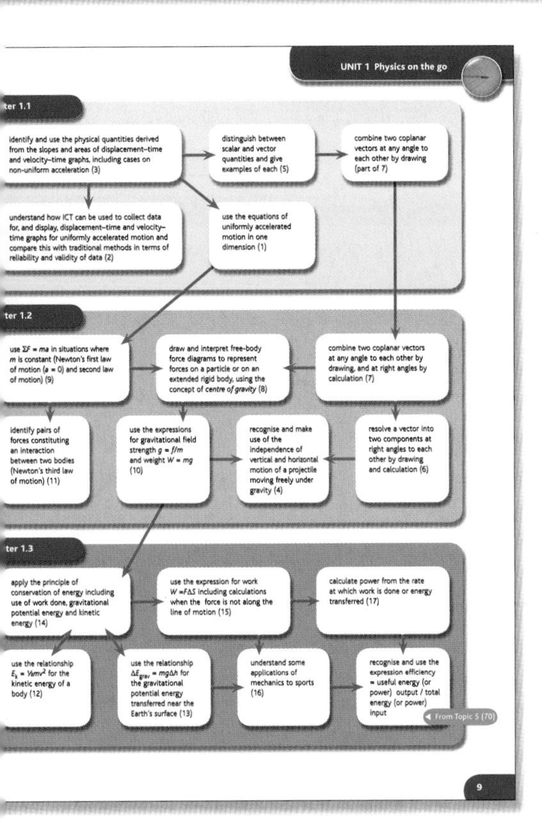

UNIT 1 Physics on the go

ter 1.1

identify and use the physical quantities derived from the slopes and areas of displacement–time and velocity–time graphs, including cases on non-uniform acceleration (3)

distinguish between scalar and vector quantities and give examples of each (5)

combine two coplanar vectors at any angle to each other by drawing (part of 7)

understand how ICT can be used to collect data for, and display, displacement–time and velocity–time graphs for uniformly accelerated motion and compare this with traditional methods in terms of reliability and validity of data (2)

use the equations of uniformly accelerated motion in one dimension (1)

ter 1.2

use $\Sigma F = ma$ in situations where m is constant (Newton's first law of motion ($a = 0$) and second law of motion) (9)

draw and interpret free-body force diagrams to represent forces on a particle or on an extended rigid body, using the concept of *centre of gravity* (8)

combine two coplanar vectors at any angle to each other by drawing, and at right angles by calculation (7)

identify pairs of forces constituting an interaction between two bodies (Newton's third law of motion) (11)

use the expressions for gravitational field strength $g = f/m$ and weight $W = mg$ (10)

recognise and make use of the independence of vertical and horizontal motion of a projectile moving freely under gravity (4)

resolve a vector into two components at right angles to each other by drawing and calculation (6)

ter 1.3

apply the principle of conservation of energy including use of work done, gravitational potential energy and kinetic energy (14)

use the expression for work $W = F\Delta S$ including calculations when the force is not along the line of motion (15)

calculate power from the rate at which work is done or energy transferred (17)

use the relationship $E_k = ½mv^2$ for the kinetic energy of a body (12)

use the relationship $\Delta E_{grav} = mg\Delta h$ for the gravitational potential energy transferred near the Earth's surface (13)

understand some applications of mechanics to sports (16)

recognise and use the expression efficiency = useful energy (or power) output / total energy (or power) input

From Topic 5 (70)

9

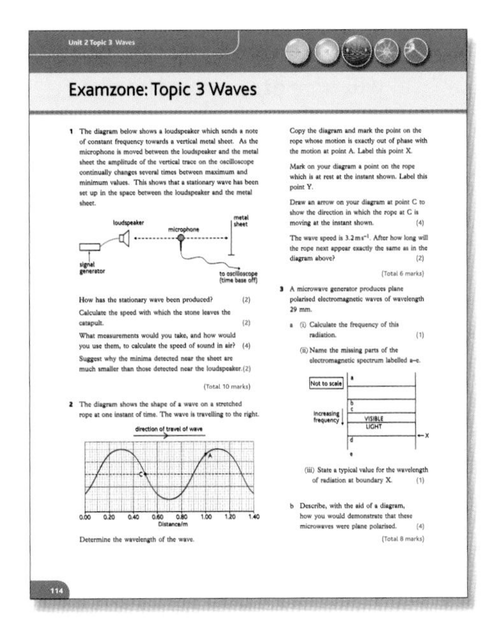

Unit 2 Topic 3 Waves

Examzone: Topic 3 Waves

1 The diagram below shows a loudspeaker which sends a note of constant frequency towards a vertical metal sheet. As the microphone is moved between the loudspeaker and the metal sheet the amplitude of the vertical trace on the oscilloscope continually changes several times between maximum and minimum values. This shows that a stationary wave has been set up in the space between the loudspeaker and the metal sheet.

How has the stationary wave been produced? (2)

Calculate the speed with which the stone leaves the catapult. (2)

What measurements would you take, and how would you use them, to calculate the speed of sound in air? (4)

Suggest why the minima detected near the sheet are much smaller than those detected near the loudspeaker. (2)

(Total 10 marks)

2 The diagram shows the shape of a wave on a stretched rope at one instant of time. The wave is travelling to the right.

Determine the wavelength of the wave.

Copy the diagram and mark the point on the rope whose motion is exactly out of phase with the motion at point A. Label this point X.

Mark on your diagram a point on the rope which is at rest at the instant shown. Label this point Y.

Draw an arrow on your diagram at point C to show the direction in which the rope at C is moving at the instant shown. (4)

The wave speed is $3.2\,m\,s^{-1}$. After how long will the rope next appear exactly the same as in the diagram above? (2)

(Total 6 marks)

3 A microwave generator produces plane polarised electromagnetic waves of wavelength 29 mm.

a (i) Calculate the frequency of this radiation. (1)

(ii) Name the missing parts of the electromagnetic spectrum labelled a–e.

(iii) State a typical value for the wavelength of radiation at boundary X. (1)

b Describe, with the aid of a diagram, how you would demonstrate that these microwaves were plane polarised. (4)

(Total 8 marks)

114

Examzone page

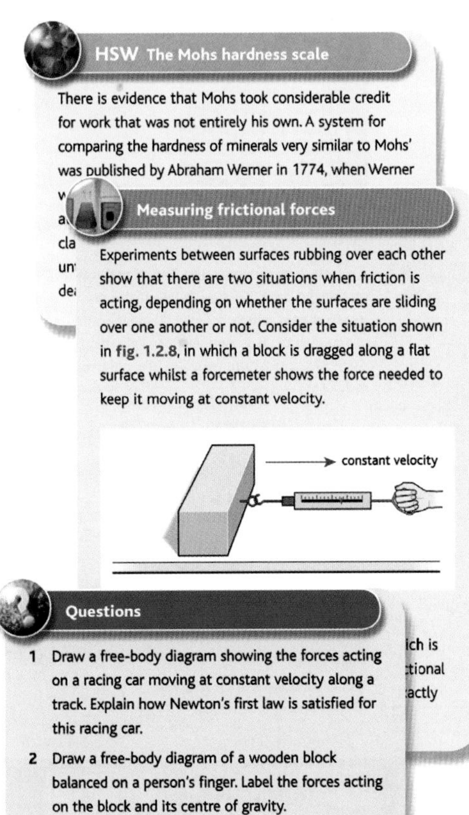

HSW The Mohs hardness scale

There is evidence that Mohs took considerable credit for work that was not entirely his own. A system for comparing the hardness of minerals very similar to Mohs' was published by Abraham Werner in 1774, when Werner

Measuring frictional forces

Experiments between surfaces rubbing over each other show that there are two situations when friction is acting, depending on whether the surfaces are sliding over one another or not. Consider the situation shown in **fig. 1.2.8**, in which a block is dragged along a flat surface whilst a forcemeter shows the force needed to keep it moving at constant velocity.

Questions

1 Draw a free-body diagram showing the forces acting on a racing car moving at constant velocity along a track. Explain how Newton's first law is satisfied for this racing car.

2 Draw a free-body diagram of a wooden block balanced on a person's finger. Label the forces acting on the block and its centre of gravity.

HSW boxes

How Science Works is a key feature of your course. The many HSW boxes within the text will help you cover all the new aspects of How Science Works that you need. These include how scientists investigate ideas and develop theories, how to evaluate data and the design of studies to test their validity and reliability, and how science affects the real world including informing decisions that need to be taken by individuals and society.

Practical boxes

Your course contains a number of core practicals that you may be tested on. These boxes indicate links to core practical work. Your teacher will give you opportunities to cover these investigations.

Question boxes

At the end of each section of text you will find a box containing questions that cover what you have just learnt. You can use these questions to help you check whether you have understood what you have just read, and whether there is anything that you need to look at again.

Examzone pages

At the end of each topic you will find two pages of exam questions from past papers. You can use these questions to test how fully you have understood the topic, as well as to help you practise for your exams.

The contents list shows you that there are two units and five topics in the book, matching the Edexcel AS specification for physics. Page numbering in the contents list, and in the index at the back of the book, will help you find what you are looking for.

How to use your ActiveBook

Find Resources

Click on this tab to see menus which list all the electronic files on the ActiveBook.

The ActiveBook is an electronic copy of the book, which you can use on a compatible computer. The CD-ROM will only play while the disc is in the computer. The ActiveBook has these features:

Student Book tab

Click this tab at the top of the screen to access the electronic version of the book.

Key words

Click on any of the words in **bold** to see a box with the word and what it means.

Interactive view

Click this button to see all the icons on the page that link to electronic files, such as documents and spreadsheets. You have access to all of the features that are useful for you to use at home on your own. If you don't want to see these links you can return to **Book view**.

UNIT 2 Topic 3 Waves

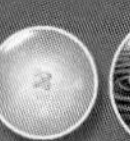

3.2 The behaviour of waves

Models of waves and their properties

Physicists use waves as models, to help them to understand why some things behave as they do. So far we have seen that stationary waves are valuable when we are trying to understand the behaviour of oscillating systems. In this chapter we shall concentrate on how waves can help us to understand the phenomena of **reflection**, **refraction**, **diffraction** and **interference**. We shall return to the subject of waves and models in Topic 5, when we consider the way in which light behaves and the explanations advanced for its behaviour.

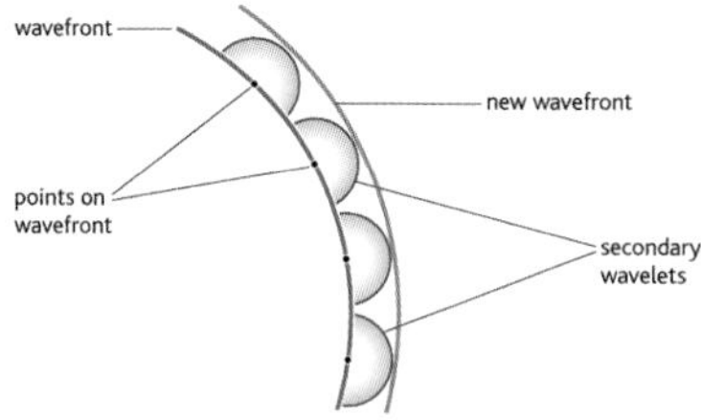

fig. 3.2.1 Huygens' construction of a wavefront. Each new wavefront has the same speed and wavelength as the original wave.

The Dutch scientist Christiaan Huygens, a contemporary of Newton, used a model of wave behaviour to explain the behaviour of waves. He explained the spreading out of a wave from a point source (like the ripple on the pond at the beginning of chapter 3.1) by considering each point on a wavefront as the source of a new set of disturbances. This representation of a wavefront is called **Huygens' construction** (fig. 3.2.1). Huygens' construction is an explanation for the way in which a circular wave spreads out, eventually leading to a plane wave as the radius of the circular wave becomes very large. This model of wave behaviour is useful in explaining other properties of waves.

Reflection

Reflection is the word used to describe what happens when a wave arrives at a barrier and changes direction. Experiments show that there is a simple relationship between the angles made with the barrier by the incident and reflected waves (fig. 3.2.3):

angle of incidence = angle of reflection

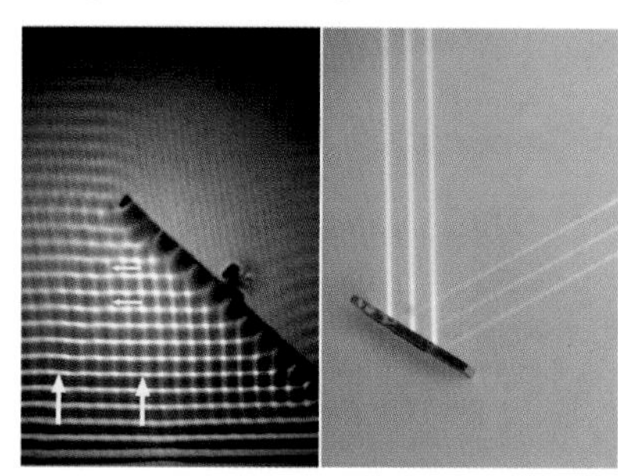

fig. 3.2.2 Reflection at a barrier of a water waves and b light. For the water waves we see the reflection of wavefronts, while for light we see the rays.

This result is known as the **law of reflection**. Notice that angles are measured between the rays and the **normal ray**, which is perpendicular to the surface of the barrier. It is important to use this convention, since the normal ray provides the only way of measuring angles where the surface is not flat.

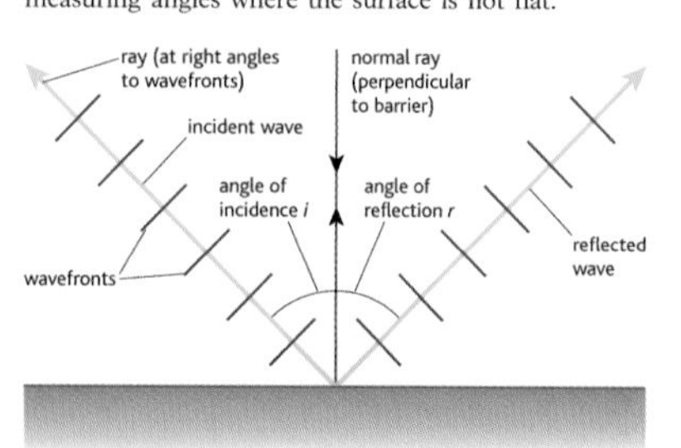

fig. 3.2.3 Reflection of waves at a barrier.

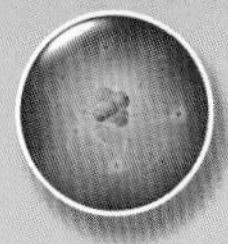

Glossary

Click this tab to see all of the key words and what they mean.

Help

Click on this tab at any time to search for help on how to use the ActiveBook.

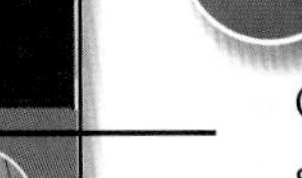

Refraction

Refraction is the change of direction of a wave that occurs when its speed changes. Refraction can be seen when light travels from one medium into another, say from air into glass or from glass into air (fig. 3.2.4a). Refraction can also be seen when water waves move from deeper water into shallower water, or vice versa (fig. 3.2.4b).

fig. 3.2.4 Refraction of a light passing through a glass block and b water waves at a change of depth. The water waves in part b are moving more slowly on the right of the photo, where the water is shallower.

Experiments with light show that there is a straightforward relationship between the angle made by the incident ray with the normal ray (the angle of incidence i) and the angle made by the refracted ray with the normal ray (the **angle of refraction** r) (fig. 3.2.5). This relationship is known as **Snell's law**, and is expressed as:

$$\frac{\sin i}{\sin r} = \text{a constant}$$

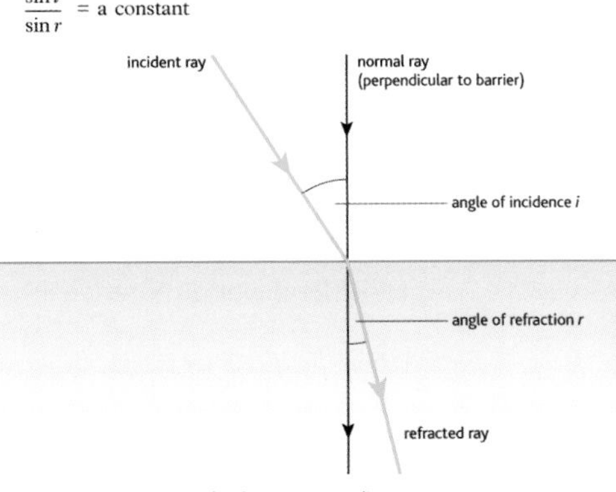

fig. 3.2.5 Refraction at a boundary between two media.

The constant is called the **refractive index** for the medium and is represented by the symbol μ, the Greek letter mu (although sometimes you will see the letter n used to represent refractive index). Table 3.2.1 shows values of refractive index for various substances.

The actual media involved in carrying the waves before and after refraction are very important. It is common to write the refractive index with subscripts indicating the medium in which the wave starts and finishes. In moving from medium 1 into medium 2, Snell's law is written as:

$$_1\mu_2 = \frac{\sin i}{\sin r}$$

Material	Crown glass	Diamond	Liquid water	Ice	Benzene	Air
Refractive index	1.52	2.42	1.33	1.31	1.50	1.0003

table 3.2.1 Values of refractive index for different materials. Since refractive index varies with wavelength for many media, the values are quoted for light with a wavelength of 5.89×10^{-7} m entering the medium from a vacuum (or air).

Zoom feature

Just click on a section of the page and it will magnify so that you can read it easily on screen. This also means that you can look closely at photos and diagrams.

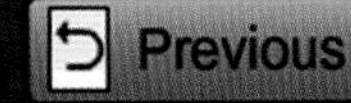
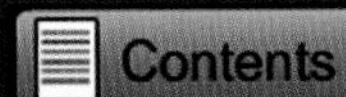

Unit 1 Physics on the go

Topic 1 Mechanics

Topic 2 Materials

Unit 2 Physics at work

TOPIC 3 Waves

TOPIC 4 DC electricity

TOPIC 5 Nature of light

Topic 1 Mechanics

This topic explains the movements of objects. It looks at how movement can be described and recorded, and then moves on to explaining why movement happens. It covers velocity and acceleration, including how to calculate these in different situations. Additionally, the effect of gravity on the movement of an object leads into consideration of the energy a body may possess or transfer.

What are the theories?

We only consider objects moving at speeds that could be encountered in everyday life. At these speeds (much less than the speed of light) Sir Isaac Newton succinctly described three laws of motion. With a knowledge of basic geometry, we can identify aspects of movement in each dimension. These three laws then allow us to calculate accurately the motion of any object over time and in three dimensions.

There are also equations for calculating kinetic energy and gravitational potential energy, and the transfer of energy when a force is used to cause the transfer. These formulae and Newton's laws can be used together to work out everything we might wish to know about the movement of any everyday object in any everyday situation.

What is the evidence?

Newton's laws of motion have been constantly under test by scientists ever since he published them in 1687. Within constraints established by Einstein in the early twentieth century, Newton's laws have always correctly described the relationships between data collected. You may have a chance to confirm Newton's laws in experiments of your own. With modern ICT recording of data, the reliability of such experiments is now much improved over traditional methods.

Whilst it is difficult for scientists to describe or identify the exact nature of energy, the equations that describe energy relationships have also consistently held up to experimental scrutiny.

What are the implications?

Combining the mathematical rules presented in this topic allows us to describe and predict the motion of all things. This statement must always be tempered by the limitations that the objects involved must be macroscopic (everyday sizes) and must be moving at reasonable speeds. Above about 10% of the speed of light, Newton's laws lose their accuracy and it becomes clear that they are, in fact, only an approximation of Einstein's more complete explanations of motion. Furthermore, if we consider subatomic particles, of which Newton knew nothing, we discover that quantum mechanics throws a probability spanner in the works.

At the end of the chapter we see the power of the equations in action as they describe the motion of the ball in a game of hockey.

The map opposite shows you all the knowledge and skills you need to have by the end of this topic. The colour in each box shows which chapter they are covered in and the numbers refer to the sections in the Edexcel specification.

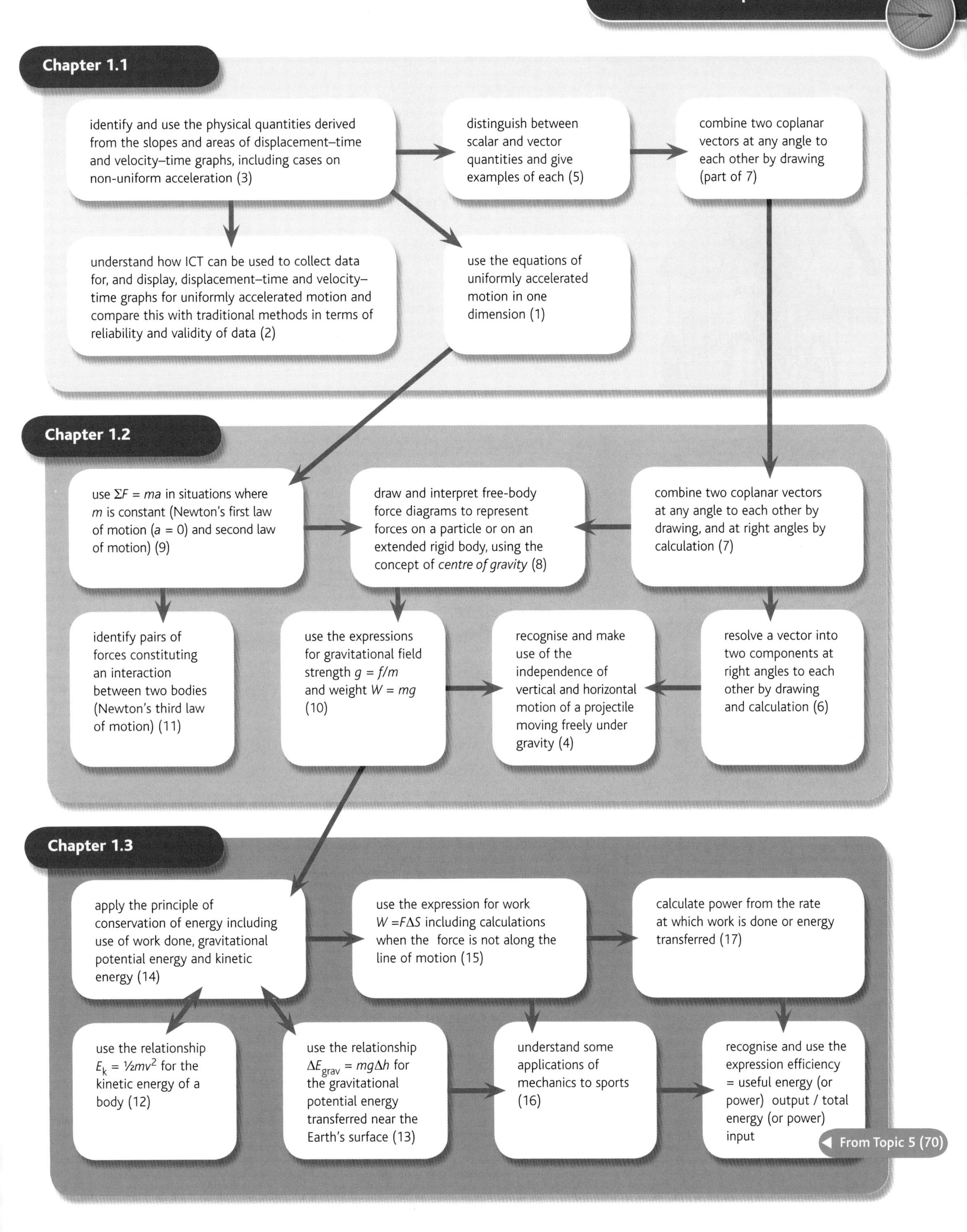
Chapter 1.1
identify and use the physical quantities derived from the slopes and areas of displacement–time and velocity–time graphs, including cases on non-uniform acceleration (3)
distinguish between scalar and vector quantities and give examples of each (5)
combine two coplanar vectors at any angle to each other by drawing (part of 7)
understand how ICT can be used to collect data for, and display, displacement–time and velocity–time graphs for uniformly accelerated motion and compare this with traditional methods in terms of reliability and validity of data (2)
use the equations of uniformly accelerated motion in one dimension (1)
Chapter 1.2
use $\Sigma F = ma$ in situations where m is constant (Newton's first law of motion ($a = 0$) and second law of motion) (9)
draw and interpret free-body force diagrams to represent forces on a particle or on an extended rigid body, using the concept of *centre of gravity* (8)
combine two coplanar vectors at any angle to each other by drawing, and at right angles by calculation (7)
identify pairs of forces constituting an interaction between two bodies (Newton's third law of motion) (11)
use the expressions for gravitational field strength $g = f/m$ and weight $W = mg$ (10)
recognise and make use of the independence of vertical and horizontal motion of a projectile moving freely under gravity (4)
resolve a vector into two components at right angles to each other by drawing and calculation (6)
Chapter 1.3
apply the principle of conservation of energy including use of work done, gravitational potential energy and kinetic energy (14)
use the expression for work $W = F\Delta S$ including calculations when the force is not along the line of motion (15)
calculate power from the rate at which work is done or energy transferred (17)
use the relationship $E_k = ½mv^2$ for the kinetic energy of a body (12)
use the relationship $\Delta E_{grav} = mg\Delta h$ for the gravitational potential energy transferred near the Earth's surface (13)
understand some applications of mechanics to sports (16)
recognise and use the expression efficiency = useful energy (or power) output / total energy (or power) input
From Topic 5 (70)

1.1 Motion

fig. 1.1.1 Average speed does not describe the speed at any particular instant.

Describing motion

Movement is a central part of our world and the Universe in which we live, whether you look for it at the scale of atoms – around 10^{-9} metres – or at the scale of our planet orbiting the Sun – around 10^{11} metres. To understand movement is to understand one of the most fundamental aspects of us and our world. This is where we shall start.

Speed and distance

Before we can attempt to begin to understand movement and what causes it, we need to be able to describe it. Let us look at the most obvious aspect of motion – **speed**.

How fast is something moving? An object's speed is calculated by dividing the distance moved by the time taken to move that distance. In the language of physics, we say that speed is *distance moved in unit time*, or:

$$\text{speed} = \frac{\text{distance}}{\text{time}}$$

So what was your speed on your way to your home yesterday? If you travelled 3 km in 15 minutes (0.25 h), your answer to this might be $12\,\text{km}\,\text{h}^{-1}$ (3 km ÷ 0.25 h). But this isn't the whole story, as **fig. 1.1.1** shows.

It is clearly unlikely that anyone will cycle at a constant speed, even without hills and stops at a shop to cope with. So the calculation of distance ÷ time in the example above tells us simply the **average speed** for the journey. It doesn't tell us anything about the speed at any given instant, as would be measured by a speedometer, for example. In fact, **instantaneous speed** is often more important than average speed. If you drive at a speed of 40 mph along a street with a 30 mph speed limit and are stopped for speeding, the police officer will not be impressed by the argument that your average speed in the last 5 minutes was only 30 mph!

It is often very useful to represent motion using a graph. This graph could plot distance against time or it could plot speed against time. **Fig. 1.1.2** shows two graphs for a journey.

O – left home
O to A – accelerating
A to B – travelling at steady speed
B to C – slowing down as going uphill
C to D – travelling at steady speed, slower than between A and B
D to E – slowing down
E to F – stopped
F to G – accelerating again
G to H – travelling at steady speed, faster than between A and B
H to J – slowing down
J – reached destination

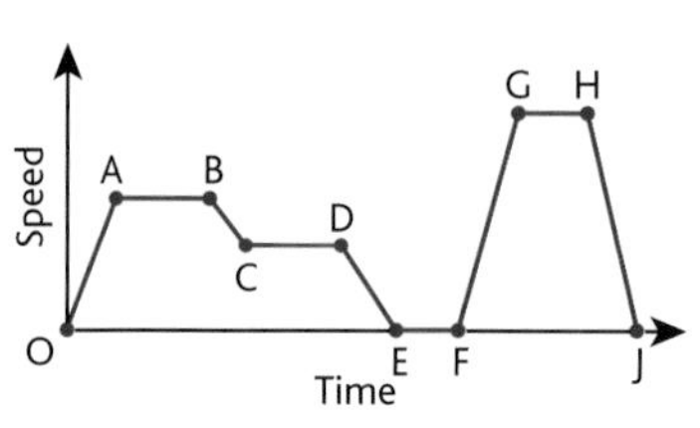

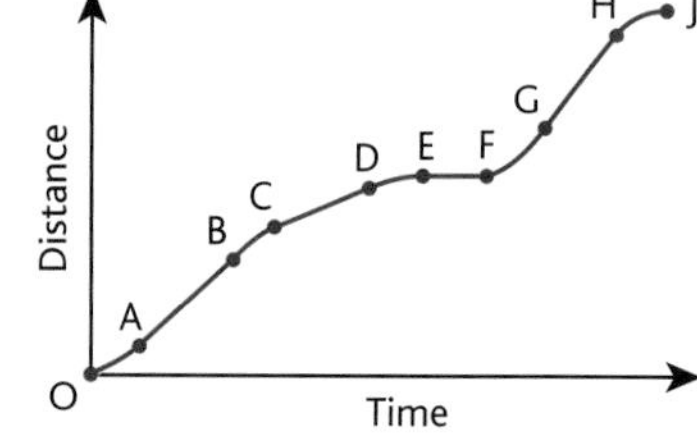

fig. 1.1.2 Speed–time and distance–time graphs for the same journey.

Look at the shapes of the two graphs very carefully. Remember that the second graph shows how far has been travelled from home, while the first graph shows the instantaneous speed of the person. Where the speed–time graph is horizontal between two points, the distance–time graph has a steady slope between the same points because the distance travelled per unit time is constant. Where the speed–time graph has a value of zero, the distance–time graph is horizontal because the person is stationary. In other words, the **slope** or **gradient** of a distance–time graph represents the speed at that particular point, the **instantaneous speed**.

Notice that steady speed corresponds to a straight line on the distance–time graph. Where an object has a steady speed, the slope of the distance–time graph is constant, and the object's average speed and its instantaneous speed are the same.

In the world of physics, speed is usually measured in metres per second ($m\,s^{-1}$) – although we are probably more used to using miles per hour (mph) in our everyday lives.

HSW Straight line graphs

Graphs are extremely useful in physics for finding and confirming relationships between different variables (for example, the stretching of a piece of metal wire and the load applied to it). The simplest type of relationship is one which is **linear**, in which a graph of one variable against another is a straight line.

Fig. 1.1.3 is an example of a linear relationship, showing how the speed of an object varies with time. Speed is plotted on the vertical axis (referred to as the 'y-axis' or the 'ordinate') and time is plotted on the horizontal axis (referred to as the 'x-axis' or the 'abscissa'). The straight line here shows that the speed of the object increases steadily with time.

The general form of the equation for a straight line is:

$$y = mx + c$$

where m is the slope or gradient of the line

and c is the **intercept** on the y-axis (the point where the line crosses the y-axis).

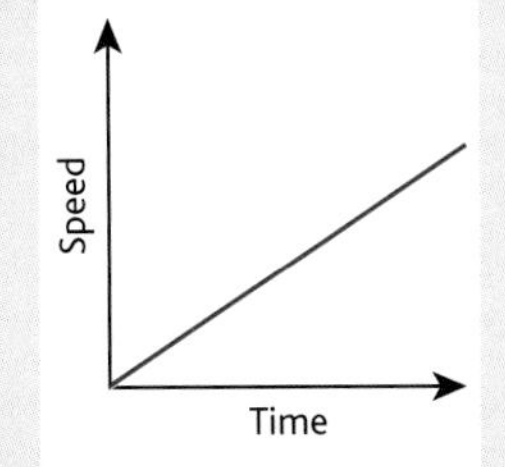

fig. 1.1.3 A speed–time graph where the speed increases steadily with time.

Recording motion

Sensors connected to a computer can be used to record the position and velocity of an object over time. With these detailed and accurate measurements, computer software can produce graphs of the data automatically.

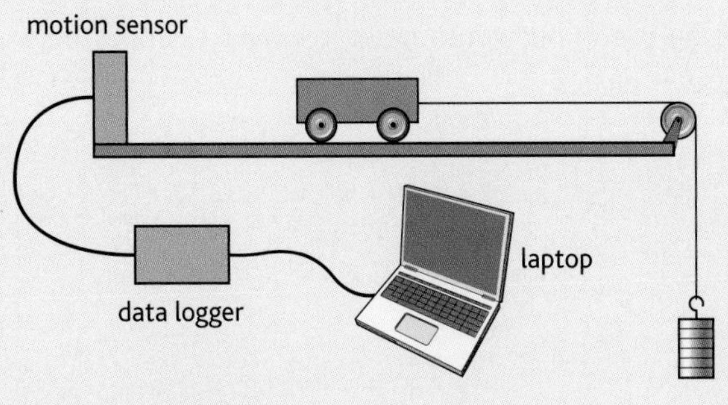

fig. 1.1.4 Data logging motion

Measurements made with electronic sensors will be more precise and more accurate than measurements made by people using stopwatches and rulers, for example. The electronic measurements will not suffer from human errors such as reaction time or misreading of scales. This means that the results and any conclusions drawn will be more reliable.

Questions

1. A horse travels a distance of 500 m in 40 s. What is its average speed over this distance?
2. Nerve impulses travel at about $100\,m\,s^{-1}$. If a woman 1.8 m tall steps on a drawing pin:
   a. roughly how long is it before she knows about it?
   b. If she is walking along with a speed of $2\,m\,s^{-1}$, how far will she have travelled in this time?
3. **Fig. 1.1.3** shows a speed–time graph for an object which starts from rest and then steadily increases speed. Sketch speed–time graphs to show the motion of an object which:
   a. has an initial speed of $5\,m\,s^{-1}$ at $t = 0$ and which then increases speed at a steady rate
   b. starts at rest at $t = 0$, stays at rest for 5 s and then increases speed at a steady rate.

Distance and displacement

Take a look at the map of the Southampton area in **fig. 1.1.5**. How far is it from Hythe to Southampton Town Quay? Of course, the distance depends on the route you take. By ferry the journey is only 1.5 miles, but by road you would need to travel 12 miles between the two places.

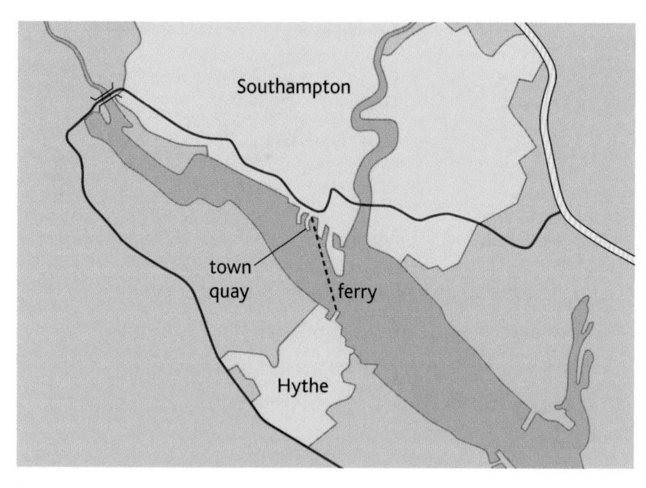

fig. 1.1.5 The distance between two places depends on the route you take. The displacement of one relative to the other does not vary.

Clearly we need a way of distinguishing between these two meanings of distance. In physics we use the terms **distance** and **displacement** to do this. If you travel from Hythe to Town Quay by road, the *distance* you have travelled is defined as the *length of path* you have taken, as measured by the mileometer in your car. However, your *displacement* is defined as the *straight line distance* between Hythe and Town Quay, as if you had taken the ferry. To describe fully the distance travelled, we only need to say how far you have gone. To describe your displacement, we not only need to specify how far you are from where you started, but also in what direction you would need to travel to get there. Distance is a **scalar** quantity – it has only size or **magnitude**. Displacement is a **vector** quantity – as well as size, it has **direction**.

Displacement and velocity

We have seen that speed is defined as distance moved in unit time. In the same way, we can now use the definition of displacement to calculate a new quantity, velocity:

$$\text{velocity} = \frac{\text{displacement}}{\text{time taken}}$$

This can also be written in symbol form as:

$$v = \frac{s}{t}$$

Like speed, velocity has magnitude. Like displacement, it also has a direction – it is a **vector**.

Speeding up and changing velocity

We are quite used to saying that an object **accelerates** as its speed increases. However, the word accelerate has a very precise meaning in physics. As we have just seen, velocity is the rate of change of displacement. Acceleration is the rate of change of velocity, so it is a vector too.

A car moving away from rest increases its speed – it accelerates. Approaching some traffic lights at red, the car slows down – this is also acceleration, but a **negative** one, because speed is taken away as the car slows down.

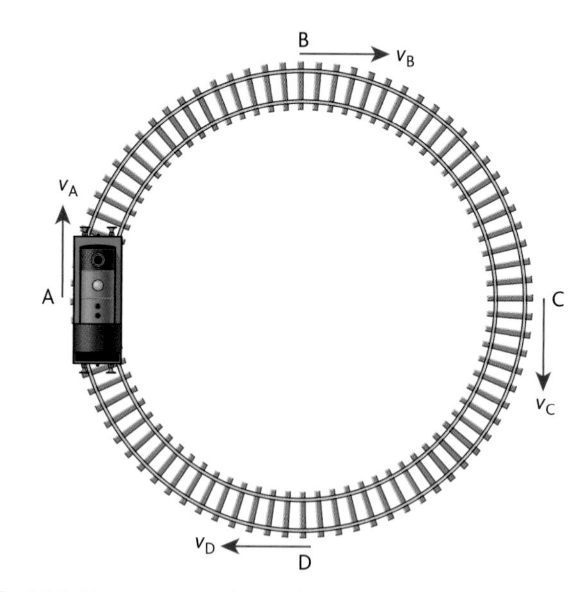

fig. 1.1.6 Movement around a circular track.

Imagine the model train in **fig. 1.1.6** moving round the track from point A through points B, C and D and back to A again at a steady speed of $0.25\,\text{m}\,\text{s}^{-1}$. Although its speed is the same at A, B, C and D its velocity is not, because the **direction** in which it is moving is different. So although its speed is constant, the train's velocity is changing – and under these circumstances we also say that it is accelerating.

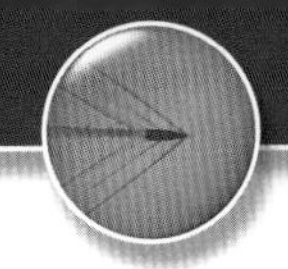

Acceleration is defined as **the rate of change of velocity with time**, and happens when there is:

a change in speed
or a change in direction
or a change in speed *and* direction.

$$\text{average acceleration} = \frac{\text{final velocity} - \text{initial velocity}}{\text{time taken for change}}$$

This can also be written in symbol form as:

$$a = \frac{v - u}{t}$$

Notice one other thing about the model train. Although its average speed between two points is always $0.25\,\text{m}\,\text{s}^{-1}$, its average velocity as it goes round the track from A and back to A again is zero, because its displacement is zero.

HSW Units

A great deal of science is based on measuring physical quantities, such as length and mass. The value of a physical quantity consists of two things – a number, combined with a unit. For example, a length may be quoted as 2.5 km or 2500 m. In order that scientists and engineers can more easily exchange ideas and data with colleagues in other countries, a common system of units is now in use in the world of science. This system is called the **Système Internationale (SI)**, and consists of a set of seven **base units,** with other **derived units** obtained by combining these. The base SI units are the metre (m), the kilogram (kg), the second (s), the ampere (A), the kelvin (K), the candela (cd) and the mole (mol). Each of these base units relates to a standard held in a laboratory somewhere in the world, against which all other measurements are effectively being compared when they are made.

fig. 1.1.7 This platinum–iridium cylinder is the standard kilogram – it is defined as having a mass of exactly 1 kg. When you buy 1 kg of apples at the supermarket you are effectively comparing their mass with the mass of this cylinder!

The units of distance, speed and acceleration show how the base units and derived units are related:

- Distance is a length, and therefore has units of **metres** in the SI system.
- Speed is distance travelled in unit time – so the units of speed (and of velocity too) are **metres per second.** This may be written as metres/second, m/s or $\text{m}\,\text{s}^{-1}$. Each of these means the same thing – metres ÷ seconds. Because $\text{s}^{-1} = 1/\text{s}$, $\text{m}\,\text{s}^{-1}$ means the same as m/s.
- Acceleration is change in velocity in unit time, and is measured in (m/s)/s – written as m/s^2 or $\text{m}\,\text{s}^{-2}$.

For vector quantities like displacement, direction must also be considered. A direction in which displacement is to be measured in a given situation is decided and displacements in this direction are then taken as positive. Velocities and accelerations then take the same sign as displacement. The choice of direction is quite arbitrary – when solving problems, the direction is usually chosen so that the mathematics involved in the solution is as simple as possible.

Questions

1 A travel brochure says that two airports are 34 km apart, and that airport A lies due south of airport B. The navigation system on board an aircraft travelling from airport A to airport B shows that it covers 380 km. Write down:

a the distance travelled by the aircraft as it flies from airport A to airport B

b the displacement of the aircraft at the end of the journey.

2 An athlete running in a sprint race crosses the finishing line and slows from a speed of $10\,\text{m}\,\text{s}^{-1}$ to rest in 4 s. What is her average acceleration?

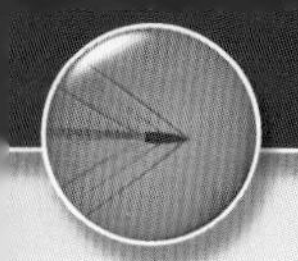

More information from graphs of motion

Velocity–time graphs are particularly useful in providing information about motion. The slope of a distance–time graph gives us information about an object's speed, because speed is rate of change of distance with time. In the same way, the slope of a velocity time graph gives us information about an object's acceleration.

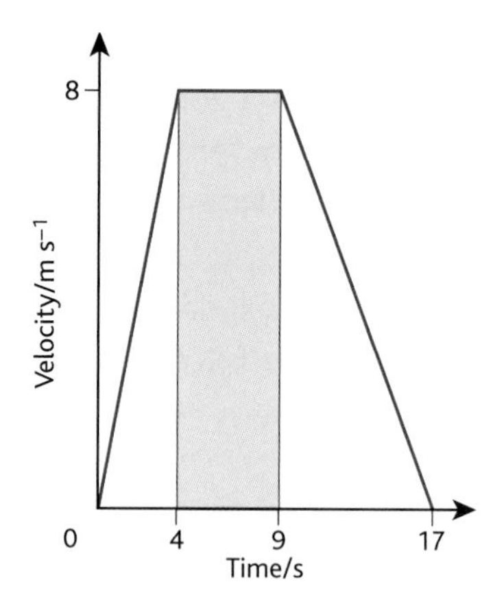

fig. 1.1.8 Velocity–time graph showing acceleration, a steady velocity, and then negative acceleration.

Look at the velocity–time graph in **fig. 1.1.8**. It tells us that:

- the object accelerates from rest to 8 m s^{-1} from 0 s to 4 s. So:

$$\text{acceleration} = \frac{\text{change in velocity}}{\text{time}}$$

$$= \frac{8 - 0}{4 - 0} = 2\,\text{m s}^{-2}$$

- there is no change of velocity from 4 s to 9 s – the acceleration is zero
- the object accelerates to rest from 9 s to 17 s. So:

$$\text{acceleration} = \frac{\text{change in velocity}}{\text{time}}$$

$$= \frac{0 - 8}{17 - 9} = -1\,\text{m s}^{-2}$$

Where a velocity–time graph is a straight line, the acceleration is uniform. Acceleration may be represented as **the rate of change of velocity with time**.

The graph in **fig. 1.1.8** also gives information about the distance travelled. Between 4 s and 9 s the object travelled with a uniform (constant) velocity of 8 m s^{-1}.

We can use this information to work out how far it travelled (its change in displacement) in this time:

$$\text{velocity} = \frac{\text{change in displacement}}{\text{time taken}}$$

so change in displacement = velocity × time taken

$$= 8 \times 5$$

$$= 40\,\text{m}$$

If you look carefully, you will see that this change in displacement represents the shaded area under the flat part of the graph. Because the area under the graph is calculated by multiplying together a velocity (in m s^{-1}) and a time (in s):

$$\frac{\text{m}}{\text{s}} \times \text{s} = \text{m}$$

the answer is in metres, and so represents a displacement.

In the same way, the area under the other parts of the graph represents displacement too:

change in displacement during initial acceleration = ½ × 8 × 4 = 16 m

change in displacement during final acceleration = ½ × 8 × 8 = 32 m

The total displacement for the whole 17 s is:

16 + 40 + 32 = 88 m

Direction of acceleration

Velocity–time graphs can give information about more complicated situations too, as **fig. 1.1.9** shows.

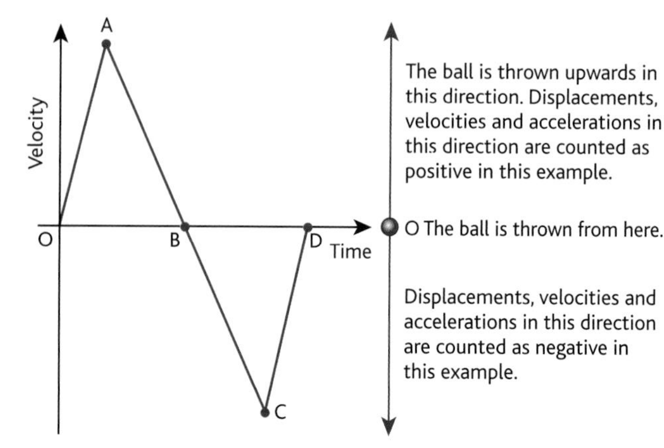

fig. 1.1.9 A highly simplified velocity–time graph for a ball being thrown upwards and then caught again.

The graph shows the motion of a ball thrown upwards, and falling back to Earth again to be caught. The ball starts from rest at time = 0. The graph is a straight line with a positive slope between O and A – this is because the person throwing the ball gives it a uniform upwards acceleration between these two points. The graph is a straight line with a negative slope between A and B – between these points the ball accelerates in a downward direction (slows down) at a steady rate, until it comes to rest at B, the highest point of its trajectory. Between B and C the graph has the same slope as it did between A and B, but its velocity is increasingly negative – it is steadily accelerating downwards (speeding up) between these points on the graph. At C the ball is caught. Between C and D the graph has a large positive slope as the person gives the ball a large upward acceleration to bring it back to rest.

Notice how the slope of each part of the graph tells us about the acceleration of the ball, while the line itself shows how the velocity of the ball changes. Careful measurement of the two areas of the graph OAB and BCD shows that they are equal, although area OAB is positive and area BCD is negative. Since the area under the line represents the ball's displacement, this shows that the ball's displacement upwards (in a positive direction) is equal to its displacement downwards (in a negative direction) – in other words, the ball falls back to Earth the same distance as it rises, and finishes up where it began.

Non-linear graphs

Although it may not be as straightforward to do, the method of measuring the area under a graph to determine the distance travelled may be used for graphs which are non-linear (not straight lines) too.

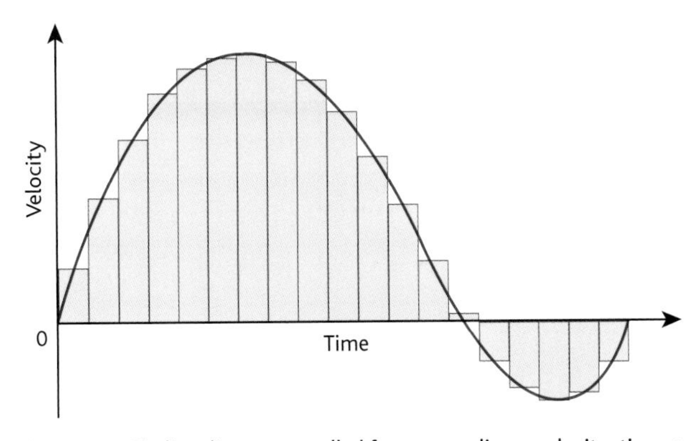

fig. 1.1.10 Finding distance travelled from a non-linear velocity–time graph.

The area under the line in **fig. 1.1.10** can be calculated by adding together the area of all the strips under the line, each of which is a rectangle. The narrower the strips, the more accurately they represent the area under the line – but the more of them there are to add up. It is important to remember to take into account whether a strip is above or below the *x*-axis when adding its area to the area of the other strips between the line and the axis.

As an example, consider the graph in **fig. 1.1.10** again. The area between the *x*-axis and the line above it is the sum of the areas of all the strips – say 350 m. (Remember the area represents a displacement.) The area between the *x*-axis and the line below it is once more the sum of the strips – say −50 m. The area is negative because it represents a displacement in the opposite direction to the first displacement. So the total displacement is the total area between the *x*-axis and the line, which is 350 m + −50 m = 300 m.

Questions

1 A train travelling along a straight track starts from rest at point A and accelerates uniformly to 20 m s^{-1} in 20 s. It travels at this speed for 60 s, then slows down uniformly to rest in 40 s at point C. It stays at rest at C for 30 s, then reverses direction, accelerating uniformly to 10 m s^{-1} in 10 s. It travels at this speed for 30 s, then slows down uniformly to rest in 10 s when it reaches point B.

a Plot a graph of the motion of the train.

b Use your graph to calculate:

i the train's displacement from point A when it reaches point C

ii the train's displacement from point A when it reaches point B

iii the train's acceleration each time its speed changes.

Moving in more than one direction – using vectors

So far we have confined ourselves to situations which are real enough, but which do not necessarily cover every type of motion found in our everyday lives. Think carefully about all the examples of motion you have seen so far and you will realise that they have all been concerned with things moving in a straight line. Whilst motion in a straight line does happen, it is usually more complex than that.

Vectors give us a fairly simple way of handling motion when it is not in a straight line. Vectors can be represented by arrows drawn to scale. The *length* of the arrow represents the **magnitude** of the vector, while the *direction* of the arrow represents the **direction** of the vector.

Combining vectors – the triangle rule

The triangle rule can be applied whenever one vector acts followed by another. For example, suppose you travel 30 m due south, and then 40 m due east – what is your displacement from your starting position?

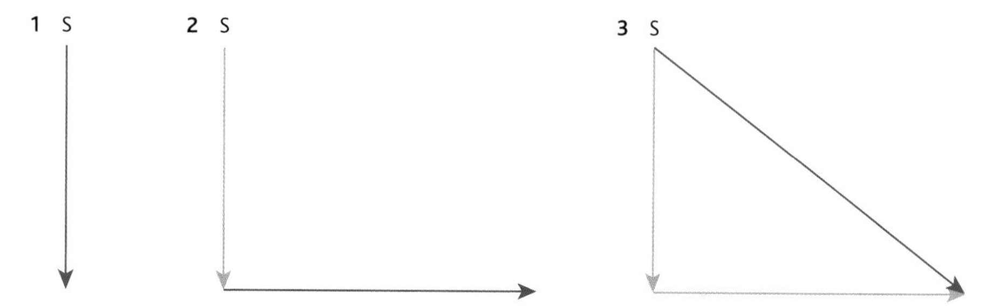

fig. 1.1.12 Adding displacement vectors using a scale diagram. 1 cm represents 10 m.

You can find your final displacement by making a scale diagram of the vectors involved. The diagrams in **fig. 1.1.12** illustrate the process:

1. Draw an arrow 3 cm long from starting point S to show a displacement of 30 m south.
2. Draw an arrow 4 cm long at right angles to the first arrow to show a displacement of 40 m east.
3. Join the starting point S to the end of the second arrow. This vector is your displacement from your starting point.

You can then measure the distance and direction of the displacement from your scale diagram. Alternatively you can use trigonometry to calculate it. In the example in **fig. 1.1.12**, the final displacement is 50 m at an angle of 53° east of the first displacement.

The sum of two or more vector quantities is called their **resultant**.

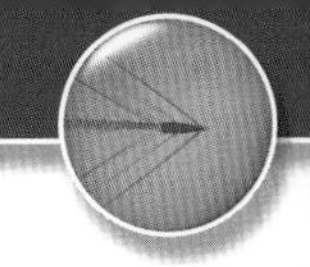

fig. 1.1.13 Getting nowhere fast! $v_{woman} + v_{walkway} = 0$

Combining vectors – the parallelogram rule

The parallelogram rule can be applied whenever vectors act at the same time or from the same point. If you have ever walked or run up a down escalator, you will have some idea of what **relative motion** is. When an object is moving, it is often very important to give some sort of information about what its motion is relative to. For example, someone running along a moving walkway may have a velocity of $2\,m\,s^{-1}$ relative to the walkway – but if the walkway has a velocity of $-2\,m\,s^{-1}$ (note the negative sign, showing that the walkway is moving in the opposite direction to the person), the person will remain in the same position relative to the ground.

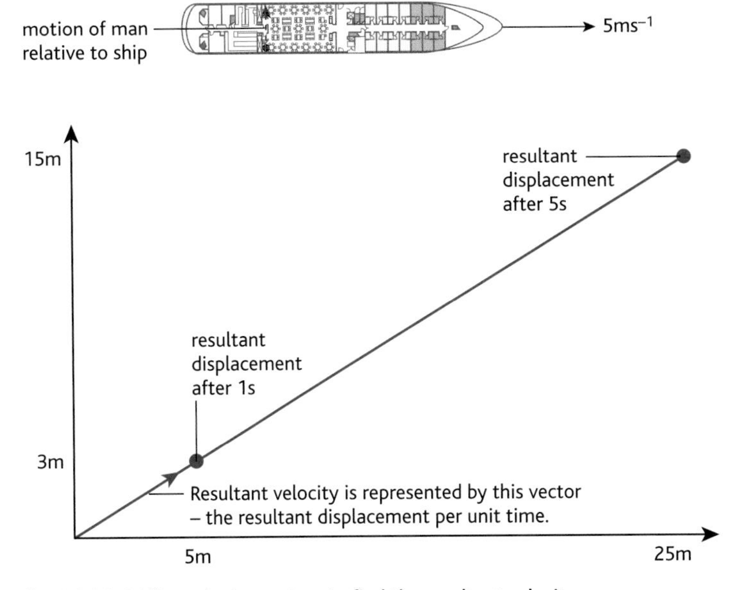

fig. 1.1.14 Adding velocity vectors to find the resultant velocity.

The resultant velocity of the woman in **fig. 1.1.13** is the sum of the vectors for her velocity relative to the walkway and the velocity of the walkway relative to the ground. Adding vectors in this type of situation, when both vectors act along the same line is easy – but a slightly different method is needed when they act along different lines.

Think about a man on a ship walking from one side of the ship to the other. If the ship is steaming forwards with a speed of $5\,m\,s^{-1}$ and the man walks from one side to the other with a speed of $3\,m\,s^{-1}$, what will be the man's movement relative to the Earth's surface?

As the man walks across the ship in **fig. 1.1.14**, the ship carries him to the right. In 1 s the man moves 3 m across the boat, and in this time the ship carries him 5 m to the right. The vector diagram shows his displacement 1 s and 5 s after starting to walk. The man's resultant velocity relative to the Earth is the vector shown. This is the resultant of the ship's velocity relative to the Earth's surface and the man's velocity relative to the ship. The resultant velocity is $5.8\,m\,s^{-1}$ in a direction making an angle of 31° with the velocity of the ship.

Questions

1. Why do aircraft take off and land into the wind?
2. A ball on a snooker table is hit by another ball and travels a distance of 50 cm due west. It is then hit again and travels a distance of 30 cm due north. Using a scale drawing, or by calculation, work out the snooker ball's displacement from its starting position.
3. A ship is travelling at $5\,m\,s^{-1}$ with a bearing of 20° east of north. There is a current of $1\,m\,s^{-1}$ flowing from the west. What is the resultant velocity of the ship?

1.2 Forces

Causes of motion

Having looked at ways of describing motion, let us look at what causes motion. The Greek philosopher Aristotle, who lived from 384 to 322 BC, said that the answer to this question was simple – motion was maintained by forces. When the force which made something move stopped acting, the object came to a standstill. In modern contexts, this idea seems quite reasonable when you think about pulling a heavy box along the floor, or pushing a car along a flat road. But what about the situation when you kick a football, for example? Once your foot ceases to be in contact with the ball, it can no longer exert a force on it, and yet the ball carries on moving for some considerable time before it eventually comes to rest.

The Italian scientist Galileo Galilei thought about problems like this, nearly 2000 years after Aristotle (Galileo was born in 1564). Galileo understood that the idea of force is central to the understanding of motion, but realised that Aristotle's explanation was incomplete. According to one story, Galileo's interest in moving objects began as a result of attending a mass in the cathedral at Pisa. During the sermon, he noticed that a cathedral lantern suspended from the roof by a long chain always took the same time to swing, whether it was swinging through a large arc or a small one. (Not having a clock, he used his own heartbeat to time the swings.) Carrying out further experiments with a pendulum, Galileo noticed that a pendulum bob always rose to very nearly the same height as it had been released from on the opposite side of its swing. Carrying this investigation further, he fixed a pin below the point of support of a simple pendulum. He raised the bob to one side and released it. The bob still rose to the height from which it was released.

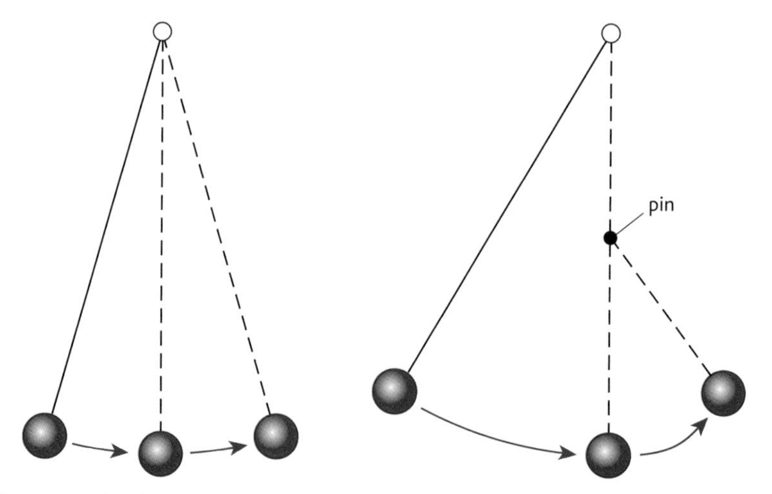

fig. 1.2.1 Galileo's pendulum experiment, which confirmed his ideas on forces and motion.

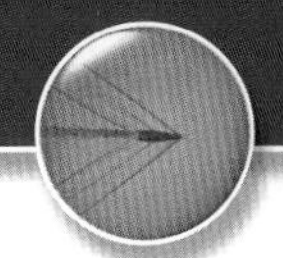

The ball and the pendulum

Galileo extended his experiment with the pendulum by carrying out a 'thought experiment', that is, one which he carried out in his head. He reasoned that if a ball rolls down a slope onto an infinitely long flat surface, by simple analogy with the pendulum experiment it will continue moving until something else causes it to stop.

Fig. 1.2.2 outlines Galileo's thought experiment.

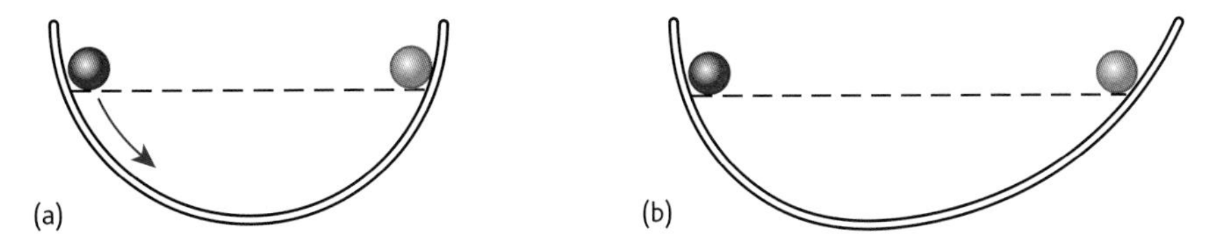

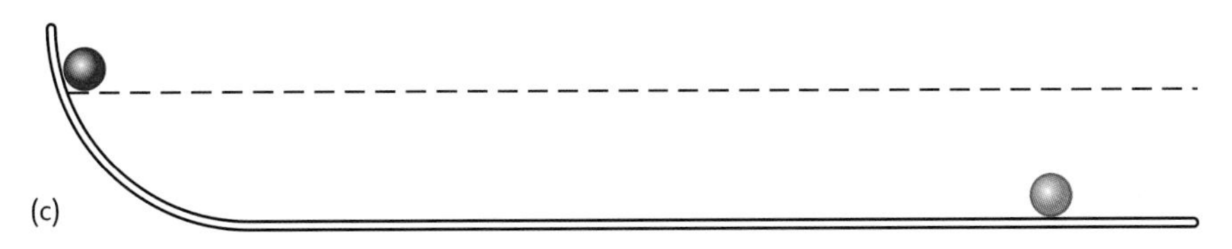

fig. 1.2.2 The diagrams show Galileo's thought experiment. The track and balls are perfectly smooth, so that there are no frictional forces between them.

By careful analogy with his pendulum experiment, Galileo reasoned that the ball would always tend to rise to the same height as it had been released from, even if it had to travel a greater horizontal distance to do so – diagrams (a) and (b) in **fig. 1.2.2** show this happening. When the rising track on the right-hand side is replaced by a flat track (c), the ball carries on moving indefinitely in an attempt to rise to its original height.

This is in direct conflict with Aristotle's explanation of the motion of objects – although it took the work of Newton to carry forward Galileo's explanation and put it on a basis that we would today recognise as being 'scientific'. Galileo had realised the importance of distinguishing between motion horizontally and vertically in a gravitational field, and had laid the foundations of the journey to the Moon, over 300 years later.

Questions

1. Aristotle argued that a force was needed in order to keep an object moving. Describe some everyday situations that are consistent with this argument. Suggest a more scientific explanation for each case that you describe.
2. 'Galileo had ... laid the foundations of the journey to the Moon.' Write a short piece for a newspaper aimed at a non-scientific audience, showing why Galileo's work was so important.

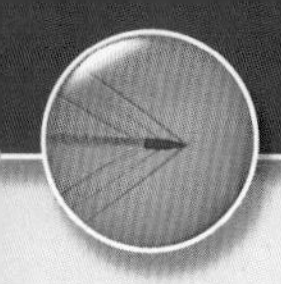

Newton's first law of motion

Forces and changing motion

The key to understanding motion is to understand forces and their interactions. The reason why we appear to need to push something to keep it moving steadily is because the motion of any object here on Earth is opposed by friction forces – and in many cases these are quite considerable. If there were no friction forces, then one push would cause an object to move indefinitely along a flat surface at a steady speed. Galileo had noticed that the concept of force was important when thinking about *changing motion* rather than motion in its own right. Galileo's work was taken up and developed by Isaac Newton, born in Lincolnshire, England in 1642, the year of Galileo's death. Building on Galileo's work, Newton framed three simple rules governing the motion of objects, which he set out as his three laws of motion in his work the *Principia*, published in 1687.

Although we now know that Newton's laws of motion break down under certain conditions (in particular, as the velocity of an object approaches the velocity of light), the laws are very nearly correct under all common circumstances.

The *Principia* was written in Latin, the language of scholarship of the time. Translated into modern English, the first law can be stated as:

> **Every object continues in its state of rest or uniform motion in a straight line unless made to change by the total force acting on it.**

In other words, an object has a constant velocity (which may be zero) until a force acts on it. So the first law of motion defines for us what a force *is*, or rather what it *does* – a force is something which can cause **acceleration**.

Newton's first law in mathematical terms

Since Newton's first law expresses motion in terms of the total force acting on a body, it can be written down involving mathematical terms. If we wish to write down 'the sum of all the forces acting on a body' we can use the mathematical expression ΣF (sigma F) to do this. So to state Newton's first law we can say:

> **If a body has a number of forces $F_1, F_2 \dots F_n$ acting on it, it will remain in a state of constant motion only if:**
>
> $\Sigma F = 0$
>
> **(that is, the sum of all the forces from F_1 to F_n is equal to zero).**

This can be calculated separately for horizontal and vertical forces. The effects of all horizontally acting forces are completely independent of those for all vertically acting forces. You will see later that this is also true for horizontal and vertical velocities, and for any pair of vectors at right angles to each other.

Free-body diagrams

Before considering the first law further, it is worth looking at how we can represent clearly the forces acting on a body.

Because a force can cause acceleration, it is a vector quantity, with both magnitude and direction. It therefore requires a way of representing both magnitude and direction on a diagram. A diagram which shows all the forces acting on a body in a certain situation is called a **free-body diagram**. A free-body diagram does not show forces acting on objects other than the one being considered.

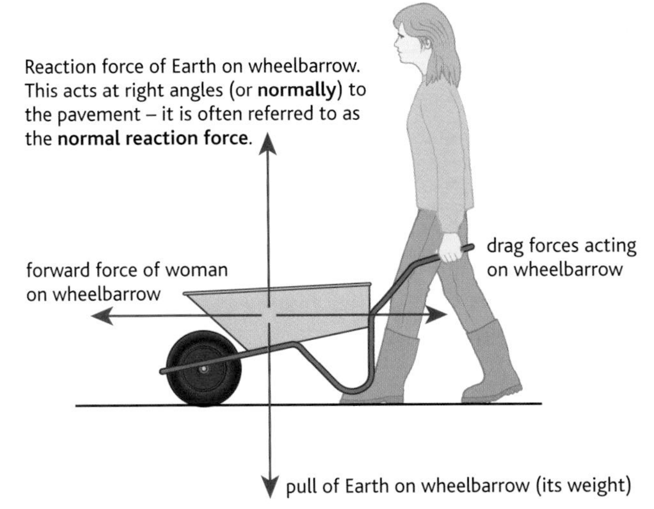

fig. 1.2.3 A simplified free-body diagram of a wheelbarrow being pushed at a steady speed along a flat surface. Notice how each force acting is cancelled out by a force exactly equal in size but opposite in direction to it. This is what Newton's first law tells us – the resultant force acting on something with constant velocity is zero.

Centre of gravity and centre of mass

In problems involving solid objects, we often draw the weight of an object as acting through a single point. This point is called the **centre of gravity**, and the justification for doing this is quite straightforward.

If we think of a ruler balanced at its midpoint, we would draw a free-body diagram of the forces acting on the ruler like that shown in **fig 1.2.4**.

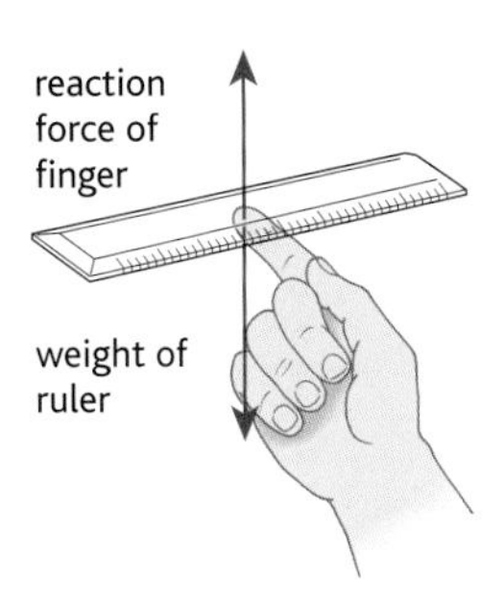

fig. 1.2.4 A balanced ruler.

This diagram assumes that we can think of the weight of the ruler as acting at its midpoint. We can justify this is by thinking of the Earth pulling vertically downwards on each particle of the ruler. As each particle on one side of the ruler has a similar particle on the other side of the ruler exactly the same distance away from the ruler's centre, the ruler will balance when it is suspended at its midpoint.

fig. 1.2.5 The centre of gravity of some uniform objects.

The centre of gravity of an object is the point at which the weight of the object appears to act. An object's **centre of mass** has a similar definition – it is the point at which all the object's mass may be considered to be concentrated. In most common circumstances (in a uniform gravitational field) an object's centre of mass and centre of gravity are at the same place, although this is not always so.

For uniform objects, the centre of mass will be at the intersection of all lines of symmetry, essentially in the middle of the object.

Questions

1 Draw a free-body diagram showing the forces acting on a racing car moving at constant velocity along a track. Explain how Newton's first law is satisfied for this racing car.

2 Draw a free-body diagram of a wooden block balanced on a person's finger. Label the forces acting on the block and its centre of gravity.

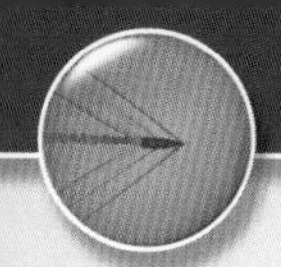

Drag forces

Once we see the situation represented in a free-body diagram like that for the wheelbarrow in **fig. 1.2.3**, it becomes quite obvious why an object stops moving when you stop pushing it. Remove the forward force acting on it and the forces on an object are no longer balanced. The resultant force now acts backwards, so the wheelbarrow accelerates backwards – that is, it slows down and eventually stops.

So why doesn't an object start moving backwards once it has stopped if there is now a resultant force acting on it? The answer to this question is because of the way that drag forces work. Drag forces in an example like the wheelbarrow are made up of two types of force – friction and air resistance – both due to matter in contact with other matter.

Where two solid surfaces rub on each other (for example in a wheel bearing or axle) **friction** always occurs. Even though they may appear perfectly smooth, the surfaces in contact are slightly rough (**fig. 1.2.6**). It is this roughness that is the cause of friction, as the two surfaces rub over one another (**fig. 1.2.7**).

fig. 1.2.6 Even the smoothest of surfaces is rough, as this high magnification photograph of a metal surface shows.

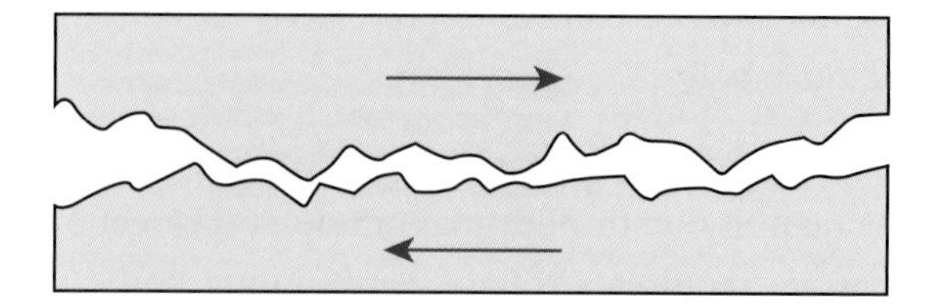

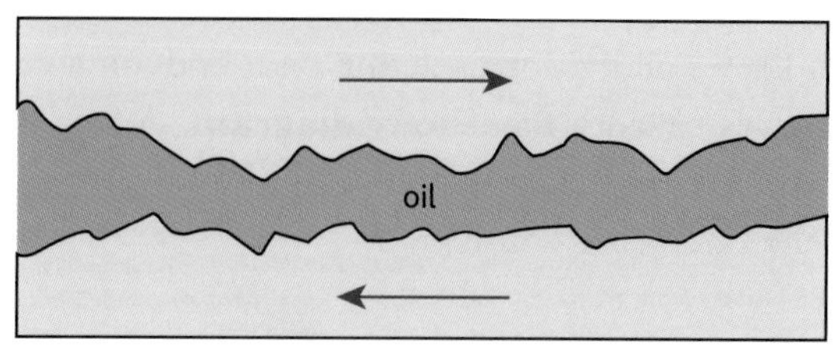

fig. 1.2.7 When two surfaces move over each other, this roughness makes it more difficult to move the surfaces – this is what we experience as friction. Oil between the surfaces pushes them apart and so reduces the frictional force between them.

This frictional force acts simply to *oppose* any motion that takes place – it cannot actually *cause* motion, as you will see if you think about the *cause* of the force. When an object comes to rest, the frictional force stops acting – it will only become important again when the object begins to move again.

Measuring frictional forces

Experiments between surfaces rubbing over each other show that there are two situations when friction is acting, depending on whether the surfaces are sliding over one another or not. Consider the situation shown in **fig. 1.2.8**, in which a block is dragged along a flat surface whilst a forcemeter shows the force needed to keep it moving at constant velocity.

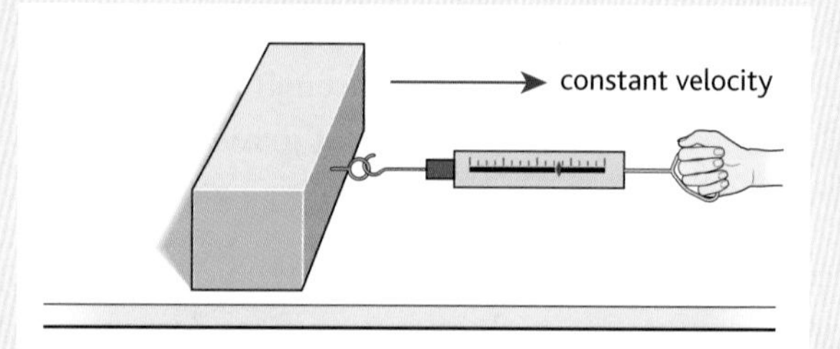

fig. 1.2.8 At constant velocity, the net force is zero.

Newton's first law tells us that, for an object which is not accelerating, $\Sigma F = 0$. This means that the frictional force resisting the motion of the box must be exactly balanced by the pulling force from the hand.

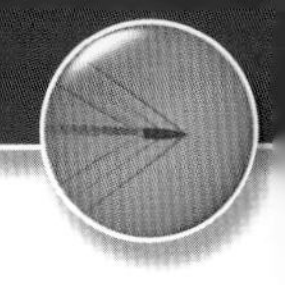

Air resistance

The other drag force which acts in this example is not important – but it is very important in many other examples. Air resistance or **aerodynamic drag** is caused when a body moves through air. In the example with the wheelbarrow, this is so small as to be insignificant.

Aerodynamic drag is caused by the fact that an object has to push air out of the way in order to move through it – and this requires a force. The force that is exerted by two surfaces rubbing together does not depend on the speed at which the two surfaces move over each other. However, the aerodynamic drag caused by an object moving through air does depend on speed – the faster the object moves, the greater the aerodynamic drag. You will learn more about this in chapter 2.1.

fig. 1.2.9 Air resistance becomes more important the faster you want to go. Careful design can reduce aerodynamic drag, by producing shapes that can 'cut through' the air and cause as little disturbance to it as possible.

Because aerodynamic drag increases as an object's velocity increases, objects with a constant driving force tend to reach a maximum velocity when they accelerate – whether they are a parachutist falling through air or a car travelling along a race track.

Free fall and terminal velocity

Someone who jumps out of a tethered balloon some way above the ground accelerates towards the ground under the influence of their own weight. They will suffer air resistance which is not insignificantly small – ask any skydiver! The acceleration with which they start to fall is called the **acceleration of free fall** or the **acceleration due to gravity**. At the surface of the Earth the value of this acceleration is $9.81\,\text{m}\,\text{s}^{-2}$. The acceleration of such a falling object is not uniform, as **fig. 1.2.10** shows.

At the top of the jump, the man is instantaneously stationary, so his air resistance is zero. The resultant force acting on him is greatest at this point, so his acceleration at this point has its maximum value.

A little later the man is moving more rapidly and his air resistance is now significant. The magnitude of his weight is still greater than his air resistance, so he is still accelerating downwards, but not as quickly as at first.

Later still his velocity has reached a point where his air resistance is equal to his weight. Now the resultant force acting on him is zero – and he is no longer accelerating. The velocity at which this happens is called the **terminal velocity**. For a human being without a parachute, terminal velocity is about $56\,\text{m}\,\text{s}^{-1}$.

On opening the parachute, the air resistance increases dramatically due to the parachute's large surface area. Now the air resistance is greater than the weight – so the resultant force on the man is upwards. The man accelerates upwards and his velocity decreases.

Eventually the man's velocity decreases to a new terminal velocity. This terminal velocity is much lower than the previous terminal velocity – about $10\,\text{m}\,\text{s}^{-1}$. Hitting the ground at this speed still requires some care – it is like jumping off a wall 5 m high!

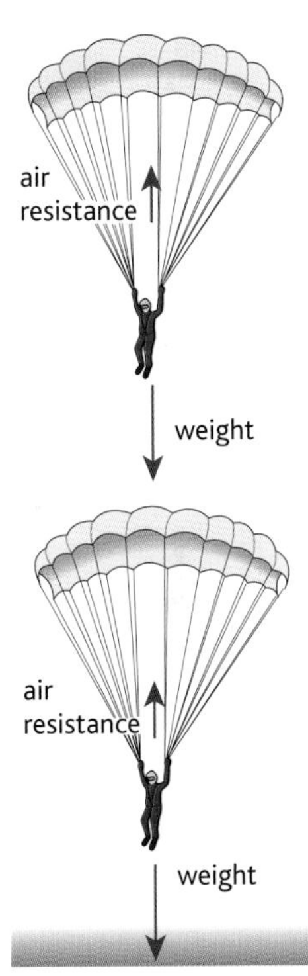

fig. 1.2.10 Free fall and terminal velocity.

Questions

1 Draw a free-body diagram showing the forces acting on a skydiver at the instant they jump from a plane.

2 Describe and explain how the resultant force on a skydiver varies from the moment they jump from a plane.

Newton's second law of motion

Having established a connection between force and acceleration which is qualitative, Newton went on to find a quantitative connection between these two. He claimed that:

$$\Sigma F = ma$$

Investigating the relationship between *F*, *m* and *a*

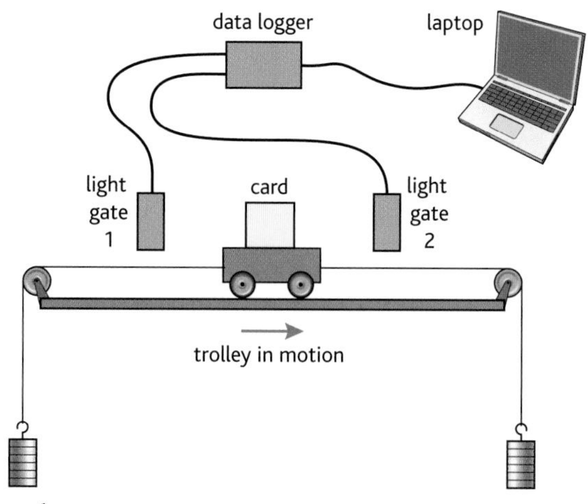

fig. 1.2.11 Experimental setup for investigating the relationship between *F*, *m* and *a*.

Using the setup shown in **fig. 1.2.11**, the acceleration can be measured for various values of the resultant force acting on the trolley while its mass is kept constant (**table 1.2.1**). By plotting a graph of acceleration against resultant force, a straight line will show that acceleration is proportional to the resultant force. A graph could also be plotted for varying masses of trolley while the resultant force is kept constant (**table 1.2.2**).

Force F/N	Acceleration a/m s^{-2}
0.1	0.20
0.2	0.40
0.3	0.60
0.4	0.80
0.5	1.00
0.6	1.20

table 1.2.1 Values of acceleration for different forces acting on a glider of mass 0.5 kg.

Mass/kg	Acceleration/m s^{-2}
0.5	1.00
0.6	0.83
0.7	0.71
0.8	0.63
0.9	0.55
1.0	0.50

table 1.2.2 Values of acceleration resulting from an applied force of 0.5 N when the mass of the glider is varied.

It is clear from **table 1.2.1** that there is a direct relationship between F and a, and that a is proportional to F (i.e. as F increases by a factor x, so does a). This can be represented as $F \propto a$.

The results in **table 1.2.2** show that there is a different relationship between a and m. Here $a \propto 1/m$ (i.e. as m increases by a factor x, a changes by a factor of $1/x$). We say that a is inversely proportional to m.

These two relationships can now be combined:

$$\left.\begin{array}{l} a \propto F \\ a \propto 1/m \end{array}\right\} \quad a \propto F/m \quad \text{or} \quad F \propto ma$$

Another way to express this is:

$$F = kma$$

where k is a constant.

By using SI units for our measurements of mass and acceleration, the units of force become kg m s^{-2} (the units of mass and acceleration multiplied together). If we **define** the unit of force in such a way that *one* unit of force accelerates a mass of *one* kilogram at a rate of *one* metre per second per second, then the constant in the equation must also have a value of *one*, and so:

$$F = ma$$

The unit of force in this system is of course better known as the **newton**. This equation defines the newton as being the resultant force which produces an acceleration of one metre per second per second when it acts on a mass of one kilogram. The mathematical statement $F = ma$ is sometimes referred to as **Newton's second law of motion**.

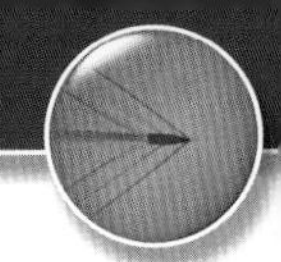

Worked examples using $F = ma$

Example 1

A runner in a sprint race reaches 9 m s^{-1} in 3 s from the start of the race. If her mass is 50 kg, what force must she exert in order to do this?

Information known:

$u = 0\,\text{m s}^{-1}$ $v = 9\,\text{m s}^{-1}$

$a = ?$ $t = 3\,\text{s}$

$s = ?$

Use equation 1:

$$v = u + at$$

Substitute values:

$$9 = 0 + a \times 3$$

$$\text{so } a = \frac{9-0}{3}$$

$$= 3\,\text{m s}^{-2}$$

Now apply $F = ma$:

$$F = 50 \times 3$$

$$= 150\,\text{N}$$

So the athlete needs to exert a force of 150 N in order to accelerate at this rate. (Will she exert this force constantly over the first 3 s of the race? Why?)

Example 2

An aeroplane lands with a velocity of 55 m s^{-1}. 'Reverse thrust' from the engines is used to slow it to a velocity of 25 m s^{-1} in a distance of 240 m. If the mass of the aeroplane is 3×10^4 kg, what is the size of the reverse thrust supplied by the engines?

Information known:

$u = 55\,\text{m s}^{-1}$ $v = 25\,\text{m s}^{-1}$

$a = ?$ $t = ?$

$s = 240\,\text{m}$

Use equation 4:

$$v^2 = u^2 + 2as$$

Substitute values:

$$(25)^2 = (55)^2 + 2 \times a \times 240$$

$$\text{so } a = \frac{625 - 3025}{2 \times 240}$$

$$= \frac{-2400}{480}$$

$$= -5\,\text{m s}^{-2}$$

Now apply $F = ma$:

$$F = 3 \times 104 \times -5$$

$$= -1.5 \times 10^5\,\text{N}$$

The reverse thrust of the engines is 150 000 N (or 150 kN). (Why was the answer obtained from the equations *negative*?)

Questions

1. a Use the results in **table 1.2.1** to plot a graph of acceleration against force.

   b Calculate the value of 1/mass for each entry in the first column of **table 1.2.2**, and plot acceleration against (1/mass) for this set of results.

   c Calculate the gradient of the best fit line for each graph.

   d What conclusions can you draw from your graphs?

2. A railway locomotive with a mass of 70 tonnes accelerates at a rate of 1 m s^{-2}. What force does the locomotive exert?

3. A 60 kg woman involved in a car accident is accelerated by her seatbelt from 14 m s^{-1} to rest in 0.15 s.

   a What is the average horizontal force acting on her?

   b How does this force compare with her weight?

Inertia, mass and weight

Newton's first law of motion is useful in considering what we mean by the term force – but it can do more too.

The tendency of an object to stay in its state of rest or uniform motion is called its inertia. Inertia is something that we all experience in our everyday lives.

- A large, massive object like a car is harder to get moving than a relatively small, light one like a bicycle.
- Without the help of a seatbelt, it can be hard for someone sitting in a moving car to stop moving when the driver of the car applies the brakes sharply.

Both of these are examples of inertia.

An object's inertia depends only on its **mass**. The definition of mass is very difficult, and you will probably have met the idea that 'mass is a measure of the amount of matter in a body'. While this statement is not false, it is not the whole truth either. The most satisfactory definition of mass uses the idea of inertia. So if two objects A and B have the same acceleration, but the resultant force on object A is $2F$ while that on object B is F, then object A must have twice the mass of object B.

Mass has only size, with no direction; it is a scalar quantity.

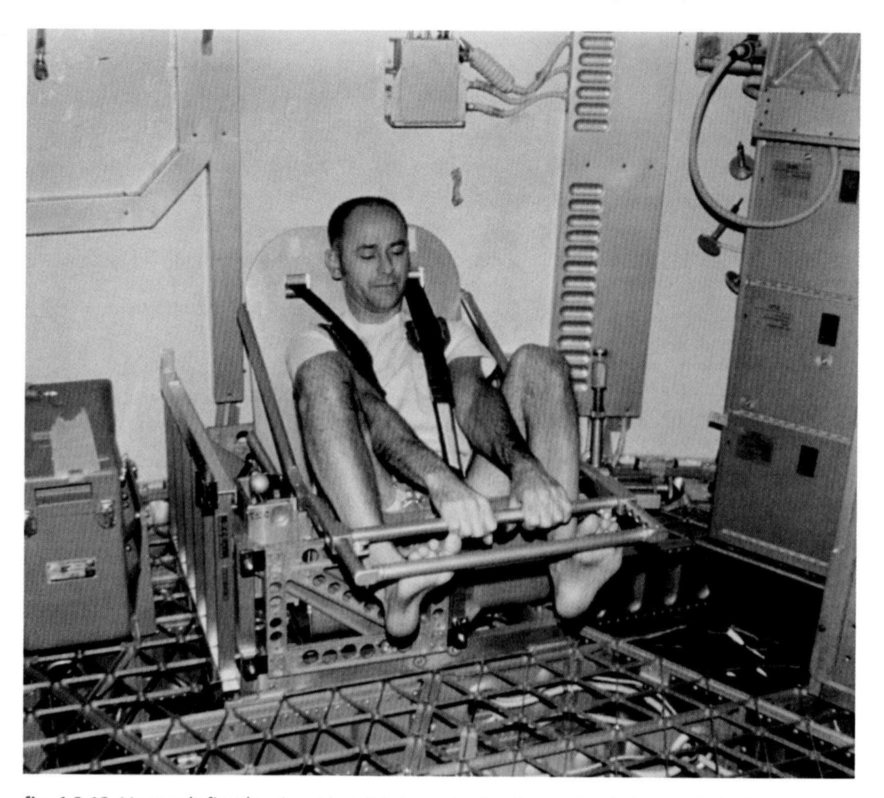

fig. 1.2.12 Newton's first law in action. It is important to know about changes in body mass happening to astronauts during long periods in orbit. Obviously bathroom scales are useless in this situation. This device uses the inertia of the astronaut's body to affect the way in which oscillations happen – the oscillations are then timed and used to calculate the mass of the astronaut.

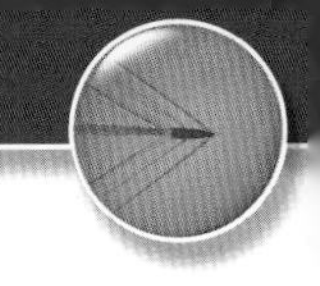

We often use the term weight in everyday life – sometimes we mean mass, rather than weight, at other times we really do mean weight. An object's weight is a force acting on it. Following Galileo's work, in Book III of the *Principia* Newton set out his theory on how masses attract one another in a process termed **gravitation**. Newton argued that it was this attractive force that we call weight. Our modern interpretation of this theory says that all masses have a **gravitational field** around them. A field is a model which physicists use to explain 'action at a distance' – the way in which two objects not in contact exert a force on each other. Using this model, a mass is said to have a gravitational field around it which causes the mass to attract another mass which is close to it. The size of the field around a particular mass depends on the size of the mass and whereabouts in the field you are.

If another mass is brought into this field, it experiences a force which pulls it towards the first mass. Weight is thus the force which is caused by gravitation – the process which occurs when one mass is brought up to another mass. The size of this force varies with the strength of the gravitational field – and as we shall see later, this varies with the position of the mass in the field. So while mass is constant no matter where it is measured, weight varies according to the strength of the gravitational field an object is in. Because weight is a force, it has both magnitude and direction – it is a vector quantity.

Gravitational field strength *g* and weight

Because the weight of an object varies according to the strength of the gravitational field it is in, this enables us to define the strength of a gravitational field.

Gravitational field strength g at a point in a gravitational field is defined as the force per unit mass acting at that point. In mathematical terms:

$$g = \frac{F}{m}$$

where F is the force acting on the object with mass m. Gravitational field strength has SI units of $\mathrm{N\,kg^{-1}}$, and is a vector quantity. The weight of an object may be calculated from this relationship, giving an expression that you will certainly have used before:

$$W = mg$$

Measuring mass and weight

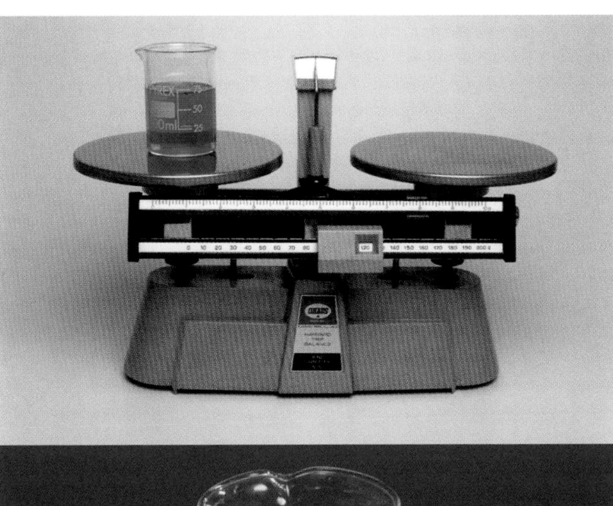

fig. 1.2.13 You might think that both these objects are being weighed. In one case you would be wrong.

The digital balance in **fig. 1.2.13** relies on a piece of conducting material being compressed or deformed by a force which changes its shape and hence its electrical resistance. The reading given therefore depends on the force an object exerts on the pan of the balance, and will be different on the Earth from the reading on the Moon – in other words, it measures the object's **weight**.

The beam balance compares the force exerted by the object on one side of the beam with the force exerted by an object of known mass on the other side of the beam. This comparison does not depend on the strength of the gravitational field that the balance is in, and the balance will give the same reading whether it is used on Earth or on the Moon – so this instrument measures the object's **mass**.

HSW Measuring *g*

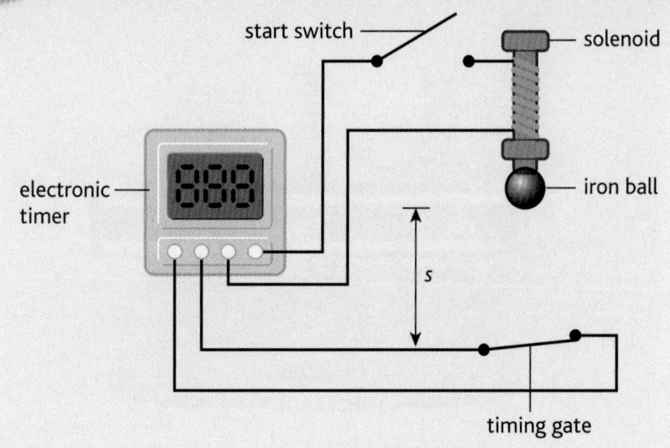

fig. 1.2.14 Apparatus for measuring g by free fall.

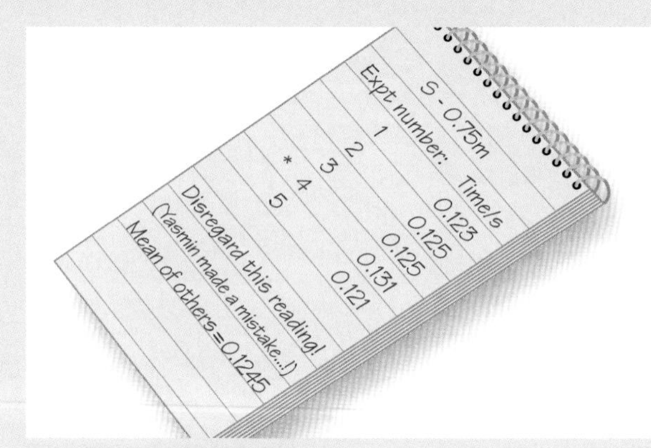

fig. 1.2.15 Free fall results

This is one way of measuring g, by analysing the free fall of an object. The iron ball is released by operating the switch, which also starts the timer. The time taken for the ball to fall the vertical distance s is measured as it passes through the timing gate. Since:

$$s = ut + \tfrac{1}{2}at^2 \quad \text{and} \quad u = 0$$

we can write:

$$s = \tfrac{1}{2}gt^2$$

This equation can be used to calculate g directly from one measurement of the falling ball. However, it is better to take *several* readings of the time to fall, take the most consistent ones, average these and then use the average to calculate g, since this should lead to a more reliable result (**fig 1.2.15**).

Even better is to notice that $s \propto t^2$. If we vary s, and plot the values of t^2 we obtain against s, the slope of the graph will be $\tfrac{1}{2}g$:

$$s = \tfrac{1}{2}gt^2$$

Compare this with $y = mx + c$ (the equation for a straight line graph). The graph will be a straight line through the origin with a gradient equal to $\tfrac{1}{2}g$.

Questions

Assume $g = 9.81\,\text{m}\,\text{s}^{-2}$.

1 A person standing on a bus is thrown towards the rear of the bus as it starts to move forwards, and to the front as it slows down. Why?

2 A person standing on the side of a ship drops a coin and sees it splash into the water 2 s later. How far above the water is the person standing?

3 An astronaut on the Moon has a weight of 128 N and a mass of 80 kg. What is the gravitational field strength on the Moon?

4 In 2002 the Canary Wharf tower in London was scaled by French urban climber, Alain Robert, using no safety devices of any kind. The top of the tower is 235 m above street level. If Robert had dropped an apple as he reached the top of the tower:

- **a** how long would it have taken the apple to fall to street level, assuming that air resistance is neglected?
- **b** assuming that he could shout loudly enough, would it be any use if Robert had shouted to warn people below? (Speed of sound in air = $340\,\text{m}\,\text{s}^{-1}$.)

Newton's third law of motion

Forces come in pairs

Our everyday experience tells us that forces come in pairs. Push a laden supermarket trolley and you can feel it pushing backwards against you. Lift a heavy bag and you can feel it pulling down on you.

Newton realised this, and stated it in his **third law of motion**. This is probably the most widely known and quoted of his laws – and it is also the most widely misunderstood! Nowadays, the law is stated as:

> **If body A exerts a force on body B, then body B exerts a force of the same size on body A, but in the opposite direction.**

It is vital to realise that these third law pairs of forces act on *different* bodies, so that a free-body diagram will only ever contain one of a given pair.

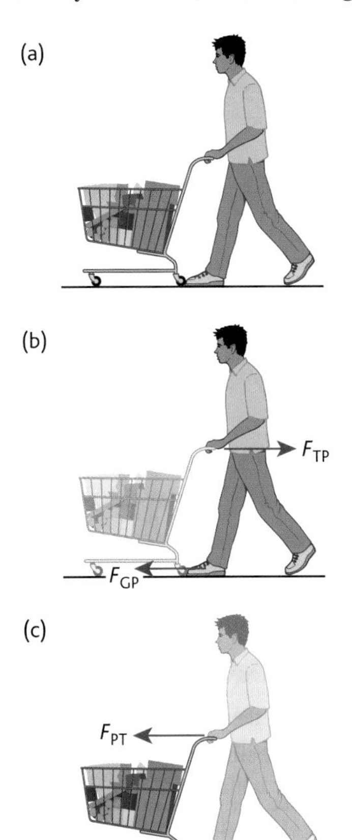

fig. 1.2.16 Think about the forces acting as you do the weekly shopping. Diagram b shows the free-body diagram for *you* as you start the trolley moving, while diagram c shows the free-body diagram for the *trolley*. Only those forces which act in a horizontal direction have been shown.

In the situation shown in **fig. 1.2.16**, only the forces F_{TP} and F_{PT} (representing the force of the trolley on the person and the force of the person on the trolley, respectively) are a third law pair. The 'missing force' is the other member of the pair to which F_{GP} (the force of the ground on the person) belongs. This would be shown on a free-body diagram for the Earth, as in **fig. 1.2.17** – and would be represented by F_{PG} using this terminology.

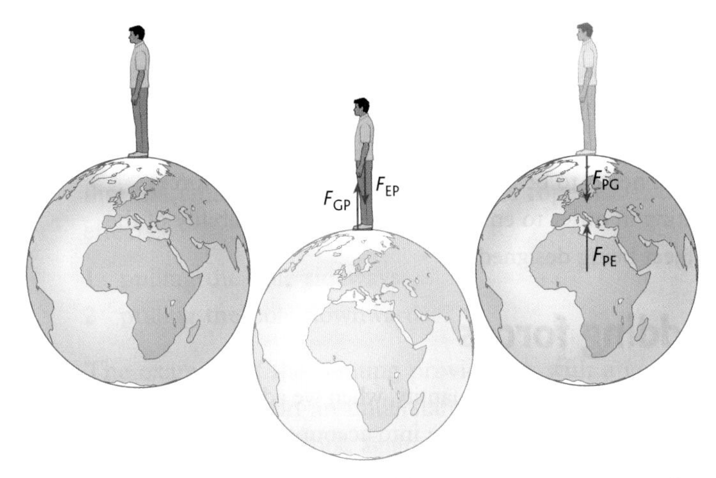

fig. 1.2.17 These diagrams show the vertical forces acting on someone standing still on the Earth.

It is tempting to assume that the push of the ground upwards on our feet is the other member of the third law pair involving the pull of the Earth downwards on us – but the free-body diagrams show that this is not so. Notice that the two third law pairs in this case are different types of force pairs. One is a gravitational pair, the other is a pair caused by contact between two surfaces. If you jump in the air, the contact pair cease to exist while you are airborne – but the gravitational pair continue to exist, to bring you (literally) back to Earth. Third law pairs of forces are always of the same type – gravitational, electrostatic, contact, etc.

Questions

1 A car is being towed by means of a rope connected to another car. Draw free-body diagrams showing the horizontal forces acting on:

a the car being towed

b the car doing the towing

c the rope.

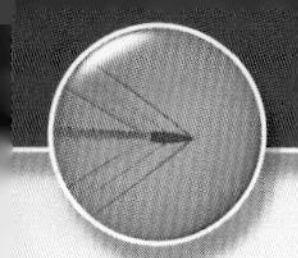

Projectiles

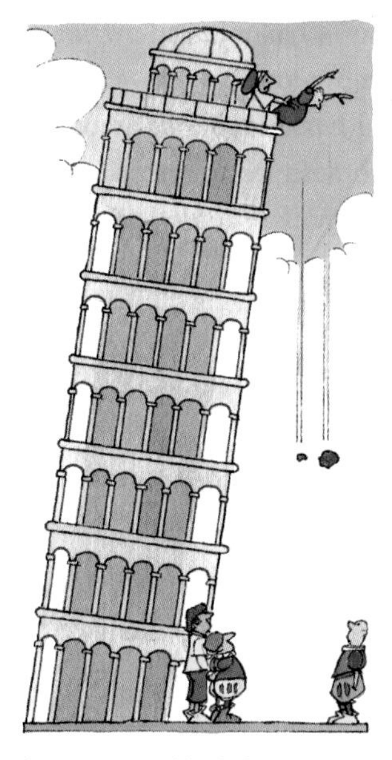

fig. 1.2.23 Galileo's famous experiment in which he dropped two unequal masses from the top of the leaning tower of Pisa. The independence of an object's mass and its acceleration in free fall was first deduced by Simon Stevinus in 1586, although this observation is usually attributed to Galileo. However, Galileo was the first to clearly state the need for the objects in question to be falling in a vacuum for this to be true rather than an approximation.

Acceleration in the Earth's gravitational field

A **projectile** is an object which is *projected* – that is to say a force acts on it to start it moving and it is then subjected to a constant force while it moves. In most cases this will mean that the object is in free fall in the Earth's gravitational field.

An object which is dropped from rest a small distance above the surface of the Earth accelerates vertically downwards under the influence of its weight. Theoretically the acceleration of an object in free fall is independent of its mass, although this is strictly true only in a vacuum – air resistance affects the motion of objects unequally, according to their cross-sectional area.

At the surface of the Earth, the rate at which an object accelerates under the influence of the Earth's gravitational field is usually known as the 'acceleration due to gravity'. The Earth's gravitational field strength and the acceleration due to gravity are usually both represented by the symbol g, although the units of the two constants are different – $\mathrm{N\,kg^{-1}}$ and $\mathrm{m\,s^{-2}}$, respectively.

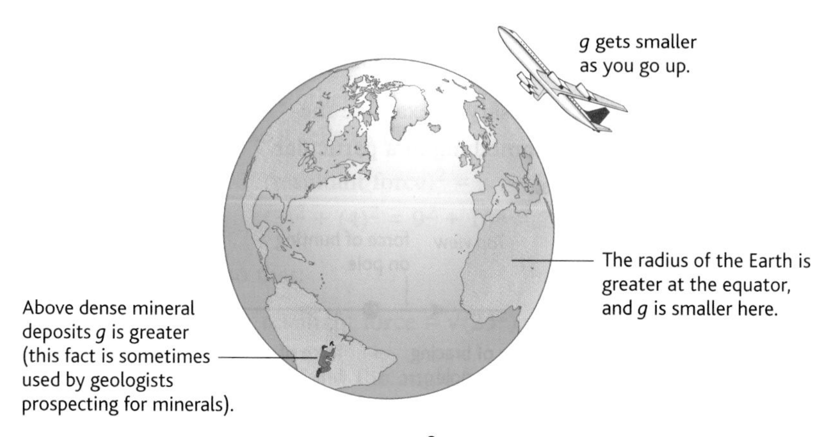

fig. 1.2.24 g is generally taken to be about $9.81\,\mathrm{m\,s^{-2}}$, although it varies at different places on the Earth's surface.

Vertical projection

fig. 1.2.25 Vertical projection.

What can the equations of motion (page 16) tell us about the motion of the ball in **fig. 1.2.25**? Taking g to be $9.81\,\mathrm{m\,s^{-2}}$ and using the information we have been given, we can write:

$u = 20\,\mathrm{m\,s^{-1}}$

$v = 0\,\mathrm{m\,s^{-1}}$ (taking the ball's velocity as zero at the top of its trajectory)

$a = -9.81\,\mathrm{m\,s^{-2}}$ $t = ?$ $s = ?$

Using equation of motion 4:

$$v^2 = u^2 + 2as$$

Substitute values:

$$(0)^2 = (20)^2 + 2 \times -9.81 \times s$$

so:

$$0 = 400 - 2 \times 9.81 \times s$$

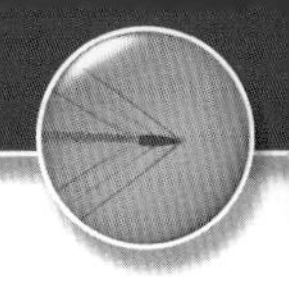

and:

$$s = \frac{400}{2 \times 9.81}$$

$$= 20.4\,\text{m}$$

The ball rises to a height of 20.4 m.

We can use equation 1 in the same way:

$$v = u + at$$

Substitute values:

$$0 = 20 + -9.81 \times t$$

so:

$$t = \frac{20}{9.81}$$

$$= 2.04\,\text{s}$$

The ball takes 2.04 s to rise to its maximum height.

Similar use of equation 2 and equation 1 should enable you to show that:

1. the ball takes 2.04 s to return to its initial height from the top of its trajectory
2. its final velocity before being caught again is $-20\,\text{m}\,\text{s}^{-1}$.

Projection at an angle

Now consider an object projected at an angle rather than vertically upwards, as shown in **fig. 1.2.26**.

The velocity of the ball can be resolved into a horizontal and a vertical component. The force diagram shows the force acting on the ball. This acts only in a vertical direction, so the ball will accelerate in a vertical direction only. As a result, the *horizontal* motion of the ball is not subject to any acceleration, and so the horizontal component of the ball's velocity is constant. In this example the horizontal velocity is $u\cos\alpha$, where α is the angle with the horizontal at which the ball is projected. The vertical component of the ball's velocity is subject to an acceleration of $-g$.

The key variable that links the vertical and horizontal motion of the ball as it travels through the air is the *time*, which is normally measured from the instant of launching the object. If we represent the horizontal displacement by the symbol s, and the vertical displacement by the symbol h, the position of the ball can be plotted as the coordinates (s, h) on a graph. The equations of motion enable us to express s and h in terms of the other variables u and α.

Using equation 2:

Horizontal motion:

$$s = (u\cos\alpha)t + \tfrac{1}{2}at^2$$

or:

$$s = ut\cos\alpha \quad \text{as} \quad a = 0$$

Vertical motion:

$$h = (u\sin\alpha)t - \tfrac{1}{2}gt^2$$ (g is in the opposite direction to h and u)

or:

$$h = ut\sin\alpha - \tfrac{1}{2}gt^2$$

The trajectory of a projectile thrown at an angle in this way is a parabola, as shown in **fig. 1.2.28**.

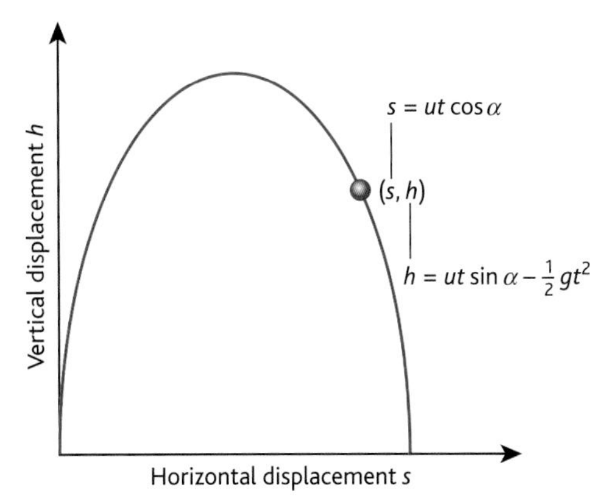

fig. 1.2.27 Trajectory of a projectile thrown at an angle.

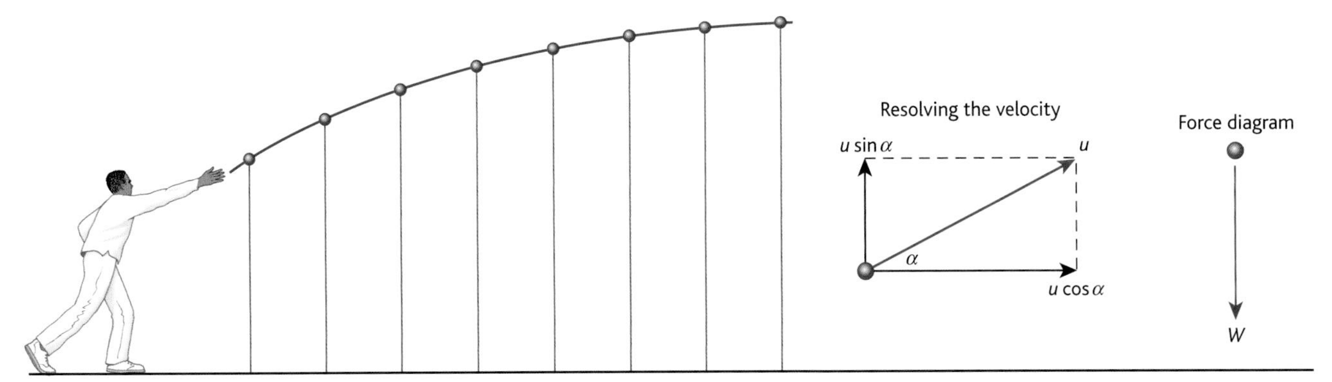

fig. 1.2.26 In each time interval the ball travels a constant horizontal distance but a varying vertical distance.

1.3 Energy and power

The concept of energy

So far we have examined the laws of motion described by Newton, and used these ideas together with those which began their development with Galileo to describe and explain the motion of objects.

The laws and rules which come from these sources are powerful tools to use in the investigation of the interactions between objects. No description of the tools that physicists use in this way would be complete without an introduction to one of the most powerful ideas that physicists have devised – the concept of **energy**.

The law of conservation of energy

We commonly think of energy as coming in a number of forms, such as chemical, electrical, kinetic and so on. Investigations of energy and its forms lead us to conclude that although energy may appear in different forms when a change occurs (for example, when a battery – containing stored chemical energy – causes an electric current – electrical energy – to flow through a wire, lighting a light bulb – producing heat energy and light energy), the total amount of energy remains constant. This rule has the status of a **law** in physics, and is called the **law of conservation of energy**. This is often expressed as:

Energy cannot be created or destroyed.

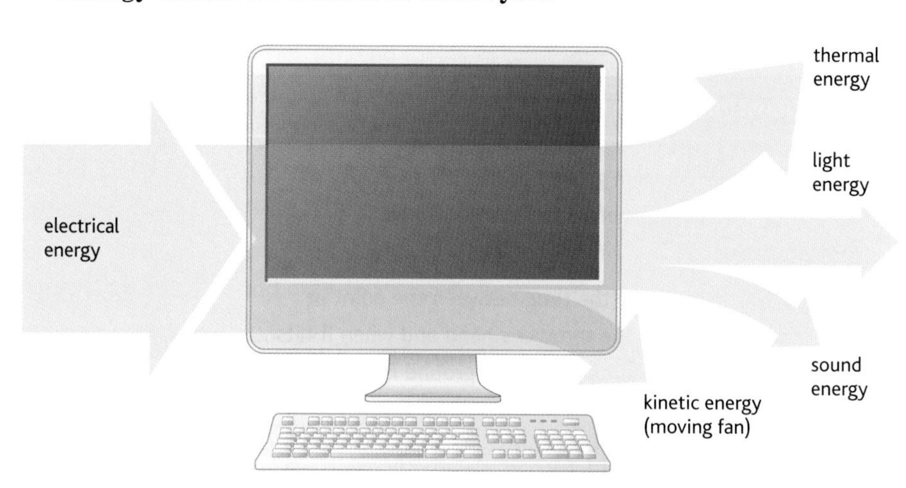

fig. 1.3.1 **The total amount of energy in a system remains constant.**

The law of conservation of energy, together with the idea of energy existing in different forms, make up an area of physics that is the most powerful and the most misunderstood of all. Essentially, the concept of energy helps us to understand and explain the way the Universe behaves – just like any other law or theory of physics.

HSW Laws and theories

At its simplest, a law of physics can be thought of as a summary of observations that physicists have made. A law is of course built on a limited number of observations, so that it is necessary for physicists to ***assume*** that it is possible to predict the behaviour of the Universe in the future based on observations made in the past. This is an assumption common to the whole of science, and the success of science as an area of study certainly seems to justify it.

A law supplies us with information about *how* the physical world may be expected to behave based on past experience, but it does not tell us *why* the physical world behaves in this way. For example, the 'energy law' we have just met says that we cannot create or destroy energy, but does not explain very much about what energy actually is or why it cannot be created or destroyed.

It is the job of **theories** in science to tell us about the behaviour of things in rather more depth. Strangely, scientists and philosophers find it very hard to agree on the way in which scientific theories come about, although most would accept that they are best described as being 'invented' (that is, made up by people) rather than simply 'discovered'.

One way in which theories are produced, based on the ideas of the philosopher Sir Karl Popper, runs as follows. A scientist wishing to explain something about the behaviour of the world will formulate a tentative idea, called at this stage a **hypothesis**. This hypothesis must be based on existing scientific knowledge, but as long as it does not contradict this knowledge, it can be as bold and imaginative as the scientist wishes. The scientist then tests the hypothesis by carrying out a number of ***experiments***. If it survives this testing, an account of the hypothesis and the tests it has survived is then published in a scientific journal so that other scientists can read about it and decide whether it is true or not – a process called peer-review. Hypotheses that survive the process of **peer-review** become accepted scientific knowledge – that is, they become theories.

It is important not to confuse use of the term 'theory' in a scientific context with its use in everyday life, where it is often used to mean 'guess'. (For example, I may have a 'theory' about why a particular football team is unlikely to win any competitions next year.) A theory in science is not a guess, but a very well-established framework that explains many observations and experimental results. A good theory will produce new hypotheses that can be tested, and so on.

fig. 1.3.2 Superstring theory suggests that we live in a Universe in which there are ten dimensions, six of which are 'coiled up' on each other. This is a very bold and imaginative theory, but it is based on far more than simple guesswork.

Questions

1. We have seen that when a driving force is removed, a moving object's drag will slow it down to a stop. This means it loses all its kinetic energy. Explain how this disappearance of energy can be in keeping with the law of conservation of energy.
2. One way of storing surplus electrical energy from a power station is to use it to raise water from a lower reservoir to a higher one. This water can then be released to generate electricity again later. (This system is referred to as 'pumped storage'.) Explain how this system might be limited by the law of conservation of energy.

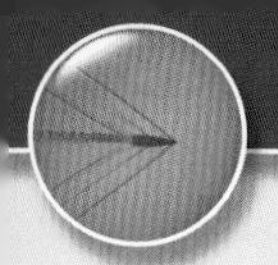

Energy transformations

The ideas behind the concept of energy represent a model for understanding the way the world (and the whole Universe) behaves. We talk about chemical energy being transferred to kinetic energy by a petrol engine. This way of talking is a quick way of saying that the petrol – because of the way the bonds in its molecules are arranged – may be burned in an engine to produce movement, and that in doing this there is a relationship between the amount of petrol used and the movement produced. We do not mean that the engine literally, as if by magic, takes something out of the petrol and uses it to turn the wheels.

fig. 1.3.3 Winding a grandfather clock.

Consider the example of winding a grandfather clock shown in **fig. 1.3.3**. The clock is driven by a falling weight. This weight can be regarded as a source of energy, in just the same way as the spring in an old-fashioned watch or the battery in a modern digital watch. Where does this energy come from in the case of the grandfather clock?

A scientific understanding of the situation (that is, one that fits with physicists' current understanding of the Universe) is that once the weight has been raised the person's muscles contain *less* of the chemicals that can be used to do something useful (like raise a weight). The raised weight, however, now has something which will let it do *more* than it could before. Energy is simply the 'accounting system' which we use to keep track of a system's ability to do something useful – rather like we use money to keep track of our ability to buy things.

An alternative explanation says that **chemical potential energy** in the person's muscles has been used to lift the weight and that, as a result, the chemical energy has been transformed into **gravitational potential energy** stored in the raised weight. This implies that something has gone from the person's muscles into the weight and has changed in some way – 'chemical energy has been transformed into potential energy'. Whilst this might *seem* to be a satisfactory explanation, it is not, because of the idea of energy transformation that it uses – a much better term to use is **energy transfer**. **Fig. 1.3.4** shows how energy transfers can be compared to transferring money.

fig. 1.3.4 When you pay a cheque into the bank you increase your ability to buy things by transferring money into your account! When you pay for something using your debit card later, you decrease it again – but no actual money changes hands in the transfer. In the same way, energy may be transferred. The energy in a system can be changed without anything actually flowing into or out of it. This is why it is better to talk about energy being *transferred* rather than to say that it is *transformed*.

Transferring energy – heating and working

Transferring energy is compared to transferring money in **fig. 1.3.4**. There are many ways of transferring money (cash, cheque, credit card, postal order, direct debit ...), but far fewer ways of transferring energy. In this section of the book we are concerned only with two of these – the ways which physicists call **heating** and **working**.

The difference between heating and working lies in the way that energy is transferred. If we heat an object, we transfer energy to it using a temperature difference – perhaps by means of a flame. If we wish to transfer energy without making use of a temperature difference, we do it by doing work – for example, by lifting an object off the floor onto a table. The terms 'heat' and 'work' therefore describe energy which has been transferred in a particular way – by means of a temperature gradient or by means of moving a force.

fig. 1.3.5 In many cases it is much easier to say 'heat energy' than 'energy transferred by means of a temperature difference' or 'work done' rather that 'energy transferred by means of moving a force'.

Work and the units of energy

As we have just seen, when energy is transferred it may be transferred by doing **work**, for example when you push a supermarket trolley across a car park. The amount of work done is calculated simply as:

work done = force × distance moved in direction of force

This simple relationship leads to the definition of the unit of energy, in the same way as Newton's second law of motion led to the definition of the unit of force. In this case, the SI unit of energy, the **joule** (J), is defined as being the energy transferred when a force of 1 newton is displaced a distance of 1 metre, that is, $1\,\text{J} = 1\,\text{N}\,\text{m}$.

Although in calculating work we are multiplying two vectors together (force and displacement), the result is a scalar. Energy has magnitude only.

If the force and the displacement are in different directions, the force must be resolved in order to calculate the work done. **Fig. 1.3.6** shows how this is done. In this figure the force is resolved so that one component ($F\cos\theta$) lies in the same direction as the displacement. This is the component of the force that is involved in transferring energy. The component of the force perpendicular to the displacement does *no* work, as it does not move in the direction in which it is acting.

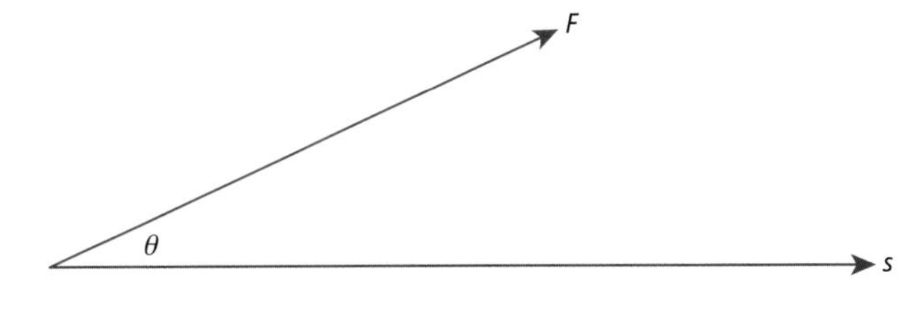

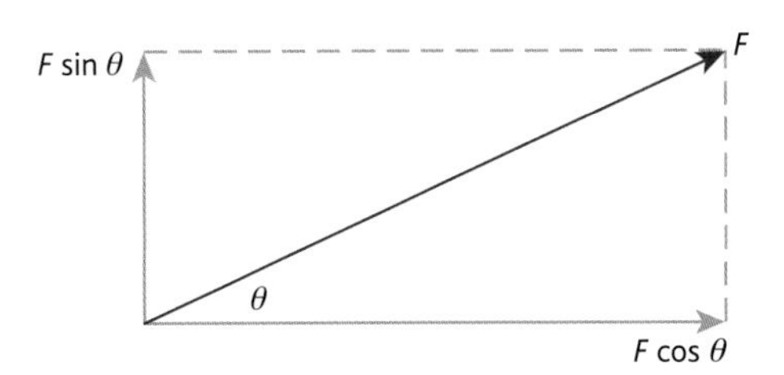

fig. 1.3.6 Resolving a force to calculate work done. In this case, work done = $F\cos\theta \times s$.

Questions

1 Give an example of a transfer of energy which could be classified as:
 a heating b working.

2 A delivery driver lifts a carton with a mass of 6.5 kg onto the back of the lorry, a height of 1.5 m from the ground. How much work is done in this energy transfer?

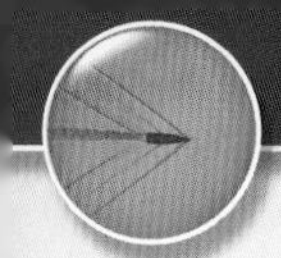

Energy and efficiency

Kinetic and potential energy

We are used to saying that an object has gravitational potential energy when it is raised through a distance Δh. In this case we write:

$$\Delta E_{grav} = m \times g \times \Delta h$$

where m is the object's mass and g is the gravitational field strength of the Earth. (Strictly, $mg\Delta h$ is the *change* in something's potential energy.)

In the same way, we talk about a moving object possessing **kinetic energy**, and write the amount of this energy as:

$$E_k = \frac{1}{2}mv^2$$

where m is the object's mass and v its velocity.

The idea of work as a means of transferring energy by moving a force lies behind both of these expressions for the energy an object has in a particular situation. The grandfather clock example (see **fig. 1.3.3** on page 40) can show us this.

When the weight in the clock is raised, a force equal and opposite to the gravitational force of attraction on the weight is applied to it. Therefore, the work done on it ('the energy transferred by means of moving a force') is equal to the weight W of the weight multiplied by the distance it is raised.

$$\begin{aligned} \text{work done} &= W \times \Delta h \\ &= mg\Delta h \\ &= \text{gravitational potential energy stored in weight } (\Delta E_{grav}) \end{aligned}$$

This potential energy may be released slowly as the weight does work on the clock mechanism ('transfers energy to the mechanism by means of a moving force'), driving the hands round.

If the wire supporting the weight breaks, the weight will fall. Instead of doing work on the clock mechanism, the weight now does work on itself (because the weight is not connected to anything, the energy transferred by the moving force has only one place to go – it stays with the weight), and its potential energy will be transferred to kinetic energy. In this case:

$$v^2 = u^2 + 2as = 2a\Delta h \quad \text{(as the weight falls from rest, } u = 0\text{, and it falls a distance } \Delta h\text{)}$$

so

$$\Delta h = \frac{v^2}{2a}$$

Now:

$$\begin{aligned} \text{kinetic energy gained} &= \text{work done on falling weight} \\ &= F \times \Delta h \\ &= F \times \frac{v^2}{2a} \end{aligned}$$

Since $F = ma$, we have:

$$\begin{aligned} \text{kinetic energy gained} &= ma \times \frac{v^2}{2a} \\ &= \frac{1}{2}mv^2 \end{aligned}$$

Pendulum energy exchange

Galileo's pendulum experiment in which he determined that the pendulum would always return to the same height illustrates the conservation of gravitational and potential energy nicely.

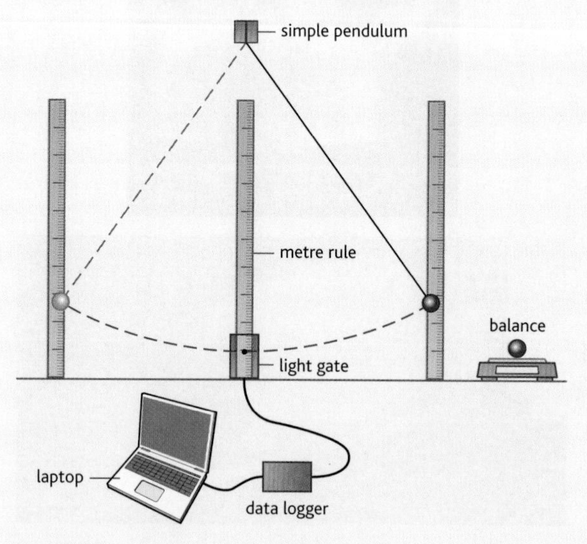

fig. 1.3.7 You can investigate energy changes for a swinging pendulum in a school laboratory.

By careful measurement of the height a pendulum rises and falls through its swing, we can determine the gravitational potential energy it loses and gains throughout one oscillation. This can then be compared with the kinetic energy it has as it passes through the lowest point. This experiment will show that the energy is constantly being transferred from kinetic to gravitational potential and back again.

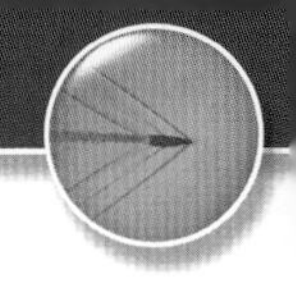

Efficiency

The idea of **efficiency** is a useful one when we are considering energy transfers. With some thought it comes as little surprise to find that it is often impossible to take all the energy in an energy store (for example a litre of petrol) and transfer it so that it does something useful. **Fig. 1.3.8** shows the energy transfers in this situation.

fig. 1.3.8 Energy transfer in a car.

Clearly this process is not 100% efficient, since in burning petrol in a car engine we want to end up with as much kinetic energy as possible – we certainly do not want thermal energy. The efficiency of this process can be calculated as:

$$\text{efficiency} = \frac{\text{useful energy got out (the kinetic energy)}}{\text{energy put in}}$$

$$= \frac{35}{100} \times 100\%$$

$$= 35\%$$

If we think of a machine or a process as a box which has energy going into it and energy coming out of it, as in **fig. 1.3.9**, then we define efficiency in the way we have just seen, that is:

$$\text{efficiency} = \frac{\text{useful energy output}}{\text{energy input}} \times 100\%$$

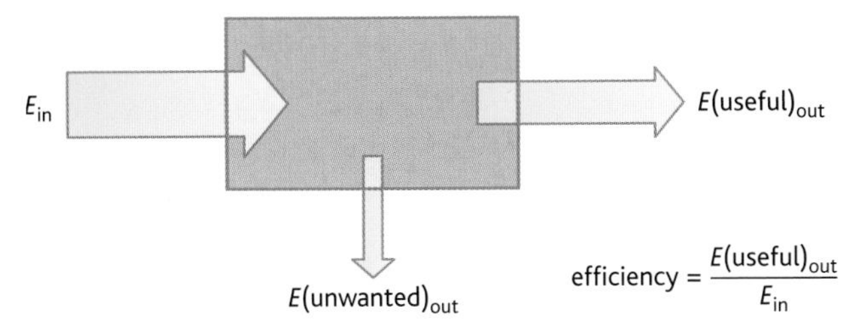

fig. 1.3.9 Energy transfer by a machine

Since efficiency is calculated by dividing one quantity in joules by another also in joules, it has no units – it is simply a ratio. It may, of course, be expressed as a percentage by multiplying the ratio by 100%.

Worked examples

A central heating boiler supplies 250 kJ of energy to hot water flowing through the boiler. In order to supply this energy to the water, the gas burned in the boiler must produce 400 kJ of energy. What is the efficiency of the boiler?

$$\text{efficiency} = \frac{\text{useful energy output}}{\text{energy input}} \times 100\%$$

$$= \frac{250 \times 10^3}{400 \times 10^3} \times 100\%$$

$$= 62.5\%$$

Questions

1 Legend has it that Galileo was only 17 years old when he started thinking about pendulum movements, whilst watching a lamp hanging on a long cable in the cathedral in Pisa, Italy. If the lamp had a mass of 1.2 kg and a draught imparted 10 J of kinetic energy to it:

  a how fast would the lamp move initially?

  b how high could the lamp rise in a swing?

2 A cricketer hits a ball straight up in the air. It leaves the bat at 16.8 m s^{-1} and has a mass of 160 g.

  a What is the kinetic energy of the ball as it leaves the bat?

  b Assuming air resistance is negligible, what is the maximum height the ball reaches above the point it left the bat?

  c If, in reality, flying up in the air and back down again is a process with an efficiency of 88%, then how fast will the ball be travelling when it returns to the start point?

Power

fig. 1.3.10 Which car has more power? The idea of power is common in everyday life. Physicists use the word 'power' to mean much the same thing as we understand when comparing the performance of cars.

In physics, the definition of power relates energy transferred to the time taken to do it. So:

$$\text{power} = \frac{\text{energy transferred}}{\text{time taken}}$$

In symbols:

$$P = \frac{E}{t}$$

Power may be measured in joules per second (J s^{-1}) in the SI system. The unit J s^{-1} is also known as the watt (W).

Since we often refer to the power developed when work is done, power may also be defined as:

$$\text{power} = \frac{\text{work done}}{\text{time taken}}$$

If a force is being moved at a steady rate, provided we know the velocity at which the force is moving and the size of the force, we can calculate the power:

$$\text{work done} = \text{force} \times \text{distance moved in direction of force}$$

and $$\text{power} = \frac{\text{work done}}{\text{time taken}}$$

so $$\text{power} = \text{force} \times \frac{\text{distance moved in direction of force}}{\text{time taken}}$$

$$= \text{force} \times \text{velocity at which force is moving}$$

In symbols:

$$P = Fv$$

Worked examples

Example 1

An athlete of mass 75 kg runs up a steep slope, rising a vertical distance of 30 m in 50 s. Neglecting the effect of any drag forces acting on the athlete, what power must his leg muscles develop in order to do this? (Assume $g = 9.81\ \text{N kg}^{-1}$.)

The athlete's muscles must supply a force to lift the athlete through a vertical distance of 30 m. Assuming that this is done at a steady rate, then the force exerted will be equal in size to the athlete's weight.

So:

$$\text{work done by athlete} = \text{weight} \times \text{vertical distance raised}$$
$$= 75 \times 9.81 \times 30$$
$$= 22\,100\ \text{J (3 s.f.)}$$

This work was done in a time of 50 s, so:

$$\text{power developed} = \frac{22\,100}{50}$$
$$= 442\ \text{W}$$

Example 2

A car towing a caravan travels along at a steady speed of $20\ \text{m s}^{-1}$. If the force exerted by the engine is 2 kN, what is the power output of the engine?

$$\text{power} = \text{force} \times \text{velocity}$$
$$= 2000 \times 20$$
$$= 40\,000\ \text{W}$$

The power output of the engine is 40 kW.

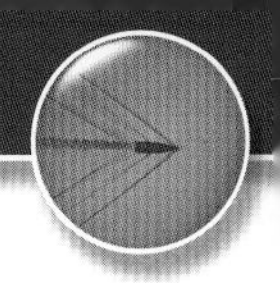

Horsepower

The power output of car engines is often expressed in **horsepower** – a method of measuring the rate of doing work dating from before the industrial revolution. $1\,\text{HP} \approx 750\,\text{W}$, so the power output of the car engine in Example 2 above is about 53 HP. In Example 1, what is the power output of the athlete's leg muscles in HP?

Investigating power and efficiency

You could investigate your own power by running up a hill of known height like the athlete in our example. A bit less tiring would be a lab investigation in which a brick is pulled up a ramp (**fig. 1.3.11**).

By pulling the brick up at a constant velocity, and using a constant force, you could calculate the power from $P = Fv$. If you also measured the time taken and the height the brick rises vertically, the efficiency of the process could be calculated.

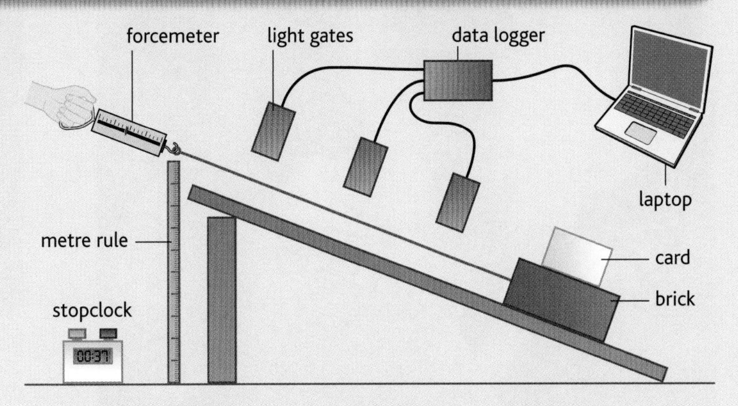

fig. 1.3.11 Experimental setup for investigating power and efficiency.

Relating power and efficiency

We saw previously that:

$$\text{efficiency} = \frac{\text{useful energy output}}{\text{energy input}} \times 100\%$$

This means that in a given time, comparing the energy output with the energy input would be the same as comparing the power output with the power input. Thus:

$$\text{efficiency} = \frac{(\text{useful energy output/time})}{(\text{energy input/time})} \times 100\%$$

Note that mathematically these two equations are the same – dividing the first equation by time/time (which is the same as dividing by one) gives us the second equation.

As $E/t = P$:

$$\text{efficiency} = \frac{\text{useful energy output/time}}{\text{power input}} \times 100\%$$

Questions

1 What is the power of a kettle which transfers 264 kJ of energy in two minutes?

2 What is the power of a luxury motorboat which is moving at a constant $22\,\text{m}\,\text{s}^{-1}$ if the total drag forces on it are 123 kN?

3 A crane lifts a steel girder with a mass of 800 kg. The girder rises 21 metres in 6 s. What is the power of the crane?

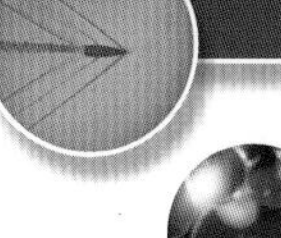

HSW The mechanics of hockey

Hockey is the second most popular sport on Earth in terms of numbers of people playing. It is a fast-moving, skilful sport in which the ball may fly at 100 miles per hour. Here we will consider how the mechanics covered in Unit 1 can apply to events in hockey.

fig. 1.3.12 Great Britain vs. Mexico

The shot at goal

If we want to calculate how fast a hockey ball would be moving after being hit hard at goal from rest, we need to think about its acceleration through Newton's second law. A standard hockey ball has a mass of 0.14 kg. If the stick applies a force of 70 N for a twentieth of a second, we can work out the answer:

$$F = ma \quad \text{so} \quad a = \frac{F}{m}$$

$$= \frac{70}{0.14}$$

$$= 500\,\text{m s}^{-2}$$

$v = u + at$ (the ball is initially stationary, so $u = 0\,\text{m s}^{-1}$, $a = 500\,\text{m s}^{-2}$, $t = 0.05\,\text{s}$)

$$= 0 + 500 \times 0.05$$

$$= 25\,\text{m s}^{-1}$$

A goal only counts if the ball is hit from inside the shooting circle. This means the ball must be no more than about 14 metres from the goal. Assuming zero drag forces, what would be the longest time the goalkeeper has to react to this shot?

fig. 1.3.13 The goalkeeper has to react quickly to save the goal.

We know the start velocity and the distance, so this is a straightforward question.

$$v = \frac{s}{t} \quad \text{so} \quad t = \frac{s}{v}$$

$$t = \frac{14}{15}$$

$$= 0.56\,\text{s}$$

The protective padding that goalkeepers wear has developed rapidly with improved materials technology. It is now much lighter and yet provides better protection than ever before. If the goalkeeper reacts quickly enough to get a leg pad in the way, how deeply will this shot squash into the foam padding which is normally around 8 cm thick? The maximum decelerating force the foam can provide to slow the ball is 1000 N.

To answer this we should consider the removal of all the ball's kinetic energy as work being done.

$$E_k = \tfrac{1}{2}mv^2$$

$$= \tfrac{1}{2} \times 0.14 \times 25^2$$

$$= 43.8\,\text{J}$$

$$W = F\Delta s$$

$$\Delta s = \frac{\Delta W}{F}$$

$$= \frac{43.8}{1000}$$

$$= 4.38 \times 10^{-2}\,\text{m}$$

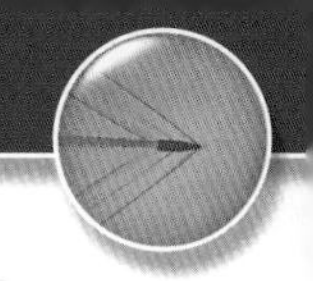

So the pad depresses by just over 4 cm. The goalkeeper is likely to feel this but it should not be painful. According to Newton's third law, when the pad exerts a force of 1000 N on the ball, the ball exerts a force of 1000 N on the pad. If the goalkeeper's leg were not protected, this force could cause a serious injury.

The aerial ball

A modern development in hockey is the aerial ball, a long pass in which a player flicks the ball high in the air in order to avoid it being intercepted. If successful, the ball may fly for four seconds and land 50 metres further downfield. To maximise the range, the player will attempt to send the ball up at an angle of 45° to the pitch surface.

With this information, we can calculate many things about the flight of this long pass. What is the initial velocity the player gives the ball; and how high does the ball travel vertically upwards?

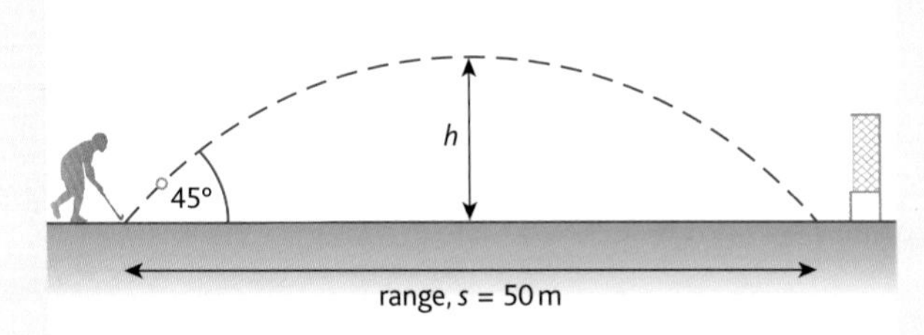

fig. 1.3.14 **Trajectory of the aerial ball.**

The horizontal velocity is easily found from the distance and time:

$$v_H = \frac{s}{t}$$

$$= \frac{50}{4} = 12.5\,\text{m s}^{-1}$$

As the angle of flight is 45° from the ground, this means the overall initial velocity can be found by trigonometry:

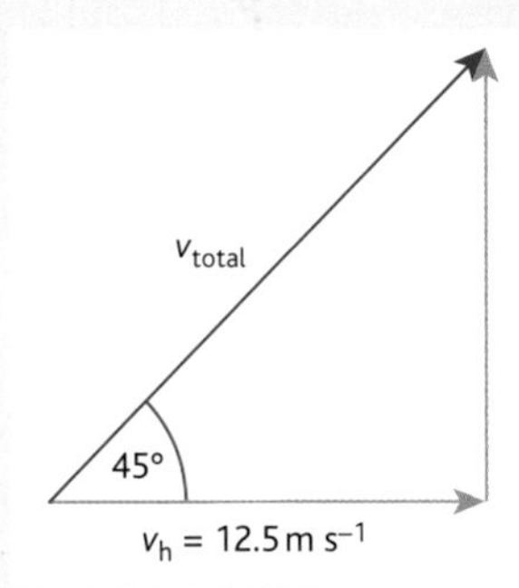

fig. 1.3.15 **Finding the horizontal component of the velocity.**

In this case:

$$\cos 45° = \frac{v_h}{v_{total}}$$

$$v_{total} = \frac{v_h}{\cos 45°}$$

$$= \frac{12.5}{0.71}$$

$$= 17.6\,\text{m s}^{-1}$$

With a 45° initial flight angle, the vertical velocity starts the same as the horizontal velocity: $v_V = 12.5\ \text{ms}^{-1}$. If we consider only the vertical motion, we can find the kinetic energy due to v_V:

$$E_k = \tfrac{1}{2} m v_v^2$$

$$= \tfrac{1}{2} \times 0.14 \times 12.5^2$$

$$= 10.9\,\text{J}$$

All the kinetic energy due to the vertical velocity of the ball is transferred into gravitational potential energy at the highest point in its flight. Remember the horizontal velocity remains constant throughout, so its contribution to the total kinetic energy is always present as kinetic energy and can be ignored in this calculation.:

$$\Delta E_{grav} = mg\Delta h$$

$$\Delta h = \frac{\Delta E_{grav}}{mg}$$

$$= \frac{10.9}{0.14 \times 9.81}$$

$$= 7.96\,\text{m}$$

Questions

1. A hockey player passes the ball at $15\,\text{m s}^{-1}$ to a team mate who stops it completely in 0.1 s. What force does the receiver's stick have to apply to the ball?
2. A penalty stroke is flicked from a distance of 6.40 m from the goal line. The striker scoops it so that the ball leaves the ground at a 45° angle and a speed of $8\,\text{m s}^{-1}$. How long does the goalkeeper have to make a save before the ball crosses the goal line?
3. The study of mechanics in sport is a popular and often profitable new area of scientific study. Describe how a sports scientist could use ICT to collect data to study the movement of players and equipment over time. Explain why technological developments have made the data collected more valid and reliable than with traditional methods of studying mechanics.

Examzone: Topic 1 Mechanics

1 A man is pushing a shopping trolley at a speed of $1.10\ \text{m s}^{-1}$ along a horizontal surface. There is a constant frictional force opposing the motion. The man stops suddenly, letting go of the trolley, which rolls on for a distance of 1.96 m before coming to rest. Show that the deceleration of the trolley is approximately $0.3\ \text{m s}^{-2}$. **(3)**

The total mass of the trolley and its contents is 28.0 kg. Calculate the frictional force opposing its motion. **(2)**

Calculate the power needed to push the trolley at a steady speed of $1.10\ \text{m s}^{-1}$. **(2)**

The man catches up with the trolley. Calculate the steady force he must now apply to it to accelerate it from rest to $1.10\ \text{m s}^{-1}$ in 0.900 s. **(3)**

(Total 10 marks)

2 A catapult fires an 80 g stone horizontally. The graph shows how the force on the stone varies with distance through which the stone is being accelerated horizontally from rest.

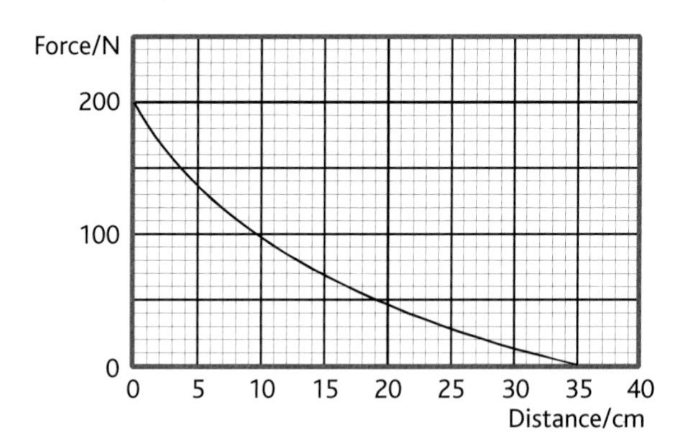

Use the graph to estimate the work done on the stone by the catapult. **(4)**

Calculate the speed with which the stone leaves the catapult. **(2)**

(Total 6 marks)

3 Two cars, A and B, are travelling along the outside lane of a motorway at a speed of $30.0\ \text{m s}^{-1}$. They are a distance d apart.

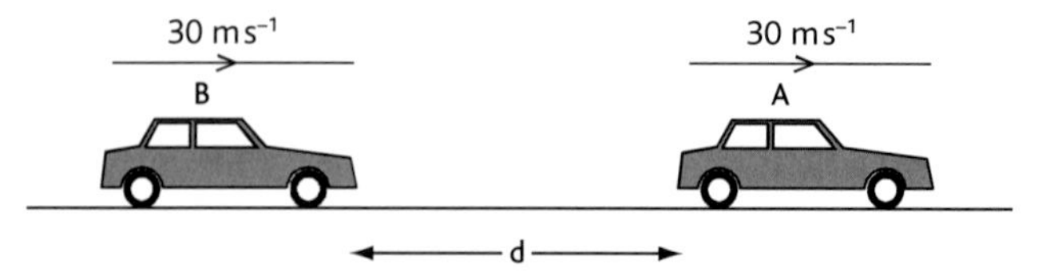

The driver of car A sees a slower vehicle move out in front of him, and brakes hard until his speed has fallen to $22.0\ \text{m s}^{-1}$. The driver of car B sees car A brake and, after a reaction time of 0.900 s, brakes with the same constant deceleration as A.

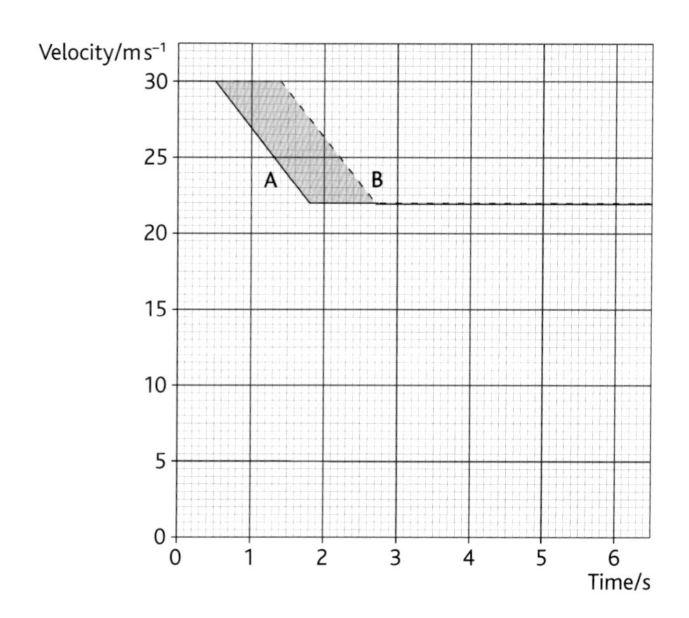

The diagram above shows velocity–time graphs for car A (solid line) and car B (broken line).

Find the deceleration of the cars whilst they are braking. **(3)**

What does the area under a velocity–time graph represent? **(1)**

Determine the shaded area. **(2)**

Suppose that, instead of only slowing down to $22.0\ \text{m s}^{-1}$, the cars had to stop. Copy the graph above and add lines to the grid above to show the velocity–time graphs in this case. (Assume that the cars come to rest with the same constant deceleration as before.) **(1)**

Explain why a collision is now more likely. **(2)**

(Total 9 marks)

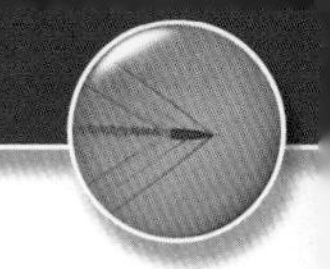

4 Explain how a body moving at constant speed can be accelerating. **(3)**

The Moon moves in a circular orbit around the Earth. The Earth provides the force which causes the Moon to accelerate. In what direction does this force act? **(1)**

There is a force which forms a Newton's third law pair with this force on the Moon.

On what body does this force act and in what direction? **(2)**

(Total 6 marks)

5 The diagram shows part of a roller coaster ride. In practice, friction and air resistance will have a significant effect on the motion of the vehicle, but you should ignore them throughout this question.

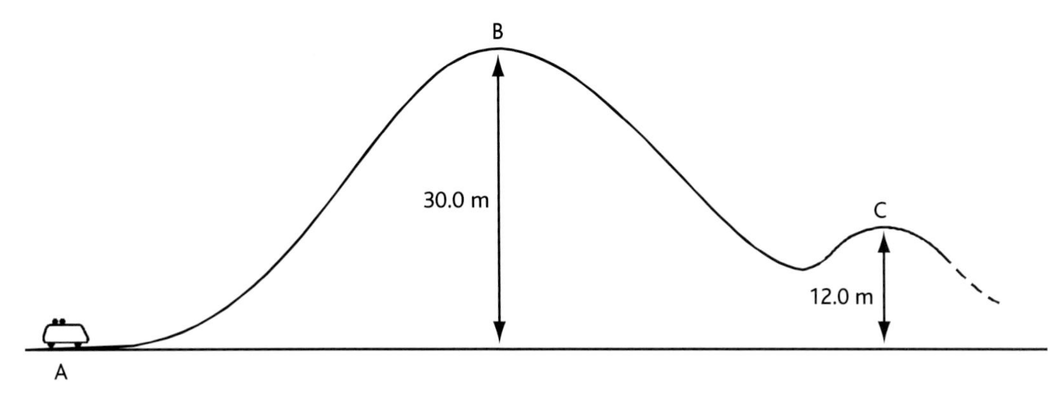

The vehicle starts from rest at A and is hauled up to B by a motor. It takes 15.0 s to reach B, at which point its speed is negligible. Copy and complete the box in the diagram below, which expresses the conservation of energy for the journey from A to B. **(1)**

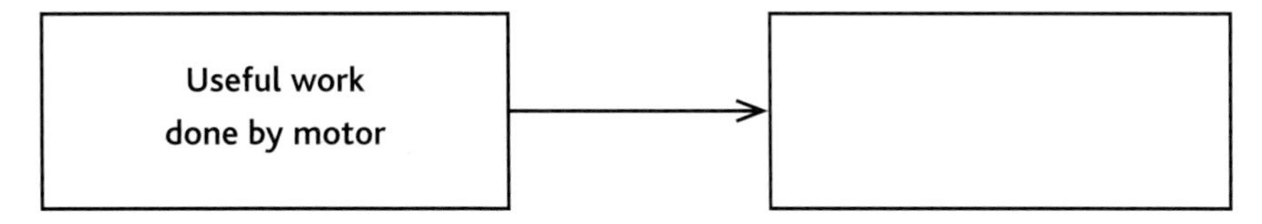

The mass of the vehicle and the passengers is 3400 kg. Calculate

(i) The useful work done by the motor.

(ii) The power output of the motor. **(4)**

At point B the motor is switched off and the vehicle moves under gravity for the rest of the ride. Describe the overall energy conversion which occurs as it travels from B to C. **(1)**

Calculate the speed of the vehicle at point C. **(3)**

On another occasion there are fewer passengers in the vehicle; hence its total mass is less than before. Its speed is again negligible at B. State with a reason how, if at all, you would expect the speed at C to differ from your previous answer. **(2)**

(Total 11 marks)

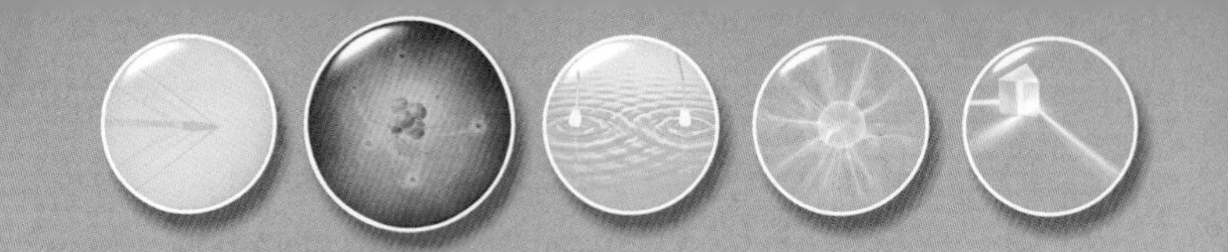

Topic 2 Materials

This topic deals with some of the macroscopic (life size) scale properties of solids, liquids and gases, and how these properties make them interact with the rest of the world around us. This is important if you want to build structures and machines that are strong enough to do their jobs.

What are the theories?

The strength of a material can be defined in many different ways, and we will look at several of these. There is a difference in the strength of a material when being crushed, compared to it being stretched to breaking point. And how does a reluctance to bend compare with a reluctance to be dented? The properties of a particular material make it suitable for some applications, but not others – so knowledge of these properties is essential for designers.

The movement of liquids and gases (fluids) is different in different conditions. Factors such as the temperature and speed of movement have a huge impact on the way that a fluid will move.

What is the evidence?

Materials testing can be done on a small scale in a school laboratory. Many of these experiments are reproduced on a larger scale for industrial applications. You will carry out some investigations of materials strength and learn how to describe the behaviour of materials. You may also have opportunities to observe the movement of fluids in simple laboratory situations.

What are the implications?

You may be able to make some strength measurements of your own – you will certainly have to make some calculations of strengths for different materials. Under various circumstances, these values can affect the design of things from tennis balls to prams to nuclear power stations.

How fluids flow is a result of properties of the liquid or gas in question. We will consider the flow of a fluid over a solid, as well as of a fluid flowing through another fluid or through itself. Understanding fluid flow is clearly important in the design of aircraft, but you may never have realised that aeronautical engineers are also employed by manufacturers of tomato ketchup.

The map opposite shows you all the knowledge and skills you need to have by the end of this topic. The colour in each box shows which chapter they are covered in and the numbers refer to the sections in the Edexcel specification.

Chapter 2.1

understand and use the terms *density*, *laminar flow*, *streamline flow*, *terminal velocity*, *turbulent flow*, *upthrust* and *viscous drag* (18)

recall and use primary or secondary data to show that the rate of flow of a liquid is related to its viscosity (19)

recognise and use the expression for Stokes' law, $F = 6\pi\eta av$ and upthrust = weight of fluid displaced (20)

investigate, using primary or secondary data, and recall that the viscosities of most fluids change with temperature. Explain the importance of this for industrial applications (21)

Chapter 2.2

investigate and use Hooke's law, $F = k\Delta x$, and know that it applies only to some materials (23)

calculate the elastic strain energy E_{el} in a deformed material sample, using the expression $E_{el} = \frac{1}{2}Fx$, and from the area under its force–extension graph (27)

explain the meaning and use of, and *calculate tensile/compressive stress*, *tensile/compressive strain*, *strength*, *breaking stress*, *stiffness* and *Young modulus*. Obtain the Young modulus for a material (24)

investigate elastic and plastic deformation of a material and distinguish between them (25)

obtain and draw force–extension, force–compression and tensile/compressive stress–strain graphs. Identify the *limit of proportionality*, *elastic limit* and *yield point* (22)

explore and explain what is meant by the terms *brittle*, *ductile*, *hard*, *malleable*, *stiff* and *tough*. Use these terms, give examples of materials exhibiting such properties and explain how these properties are used in a variety of applications (26)

2.1 Fluid flow

Fluids

Have you ever wondered why it is sometimes so difficult to get tomato ketchup out of the bottle? The answer is that the manufacturers make it thick on purpose. Market research shows that consumers enjoy a certain consistency of ketchup on their chips, and producing it that thick makes the sauce flow very slowly.

This chapter will explain various aspects of the movements of fluids, including some of the ways in which fluid properties are measured. A **fluid** is defined as any substance that can flow. Normally this means any gas or liquid, but solids made up of tiny particles can sometimes behave as fluids. An example is the flow of sand through an hourglass.

Density

One of the key properties of a fluid is its **density**. Density is a measure of the mass per unit volume of a substance. Its value depends on the mass of the particles from which the substance is made, and how closely those particles are packed:

$$\text{density, } \rho \ (\text{kg m}^{-3}) = \frac{\text{mass, } m \ (\text{kg})}{\text{volume, } V \ (\text{m}^3)}$$

The equation for calculating density works for mixtures and pure substances, and for all states of matter. Thus, fluid density is also mass per unit volume.

fig. 2.1.1 Density is very important in determining how heavy an object is.

Worked examples

Example 1

A house brick is 23 cm long, 10 cm wide and 7 cm high. Its mass is 3.38 kg.

What is the brick's density?

$$\rho = \frac{m}{V}$$

volume, $V = 0.23 \times 0.10 \times 0.07 = 1.61 \times 10^{-3}\ \text{m}^3$

mass, $m = 3.38\ \text{kg}$

$$\text{density, } \rho = \frac{3.38}{1.61 \times 10^{-3}}$$

$\rho = 2100\ \text{kg m}^{-3}$

Example 2

At 20 °C, a child's balloon filled with helium is a sphere with a radius of 20 cm. The mass of helium in the balloon is 6 grams. What is the density of helium at this temperature?

$$\rho = \frac{m}{V}$$

volume, $V = (4/3)\pi r^3 \qquad r = 0.20\ \text{m}$

$= (4/3)\pi\ (0.20)^3$

$= 0.0335\ \text{m}^3$

mass, $m = 0.006\ \text{kg}$

$$\text{density, } \rho = \frac{0.006}{0.0335}$$

$\rho = 0.179 \approx 0.18\ \text{kg m}^{-3}$

Material	State	Density/kg m^{-3}
Air	Gas (sea level, 20 °C)	1.2
Pure water	Liquid (4 °C)	1000
Sulfuric acid (95% conc)	Liquid (20 °C)	1839
Cork	Solid	240
Ice	Solid	919
Window glass	Solid	2579
Iron	Solid	7850
Gold	Solid	19 320

table 2.1.1 Examples of density values for solids, liquids and gases.

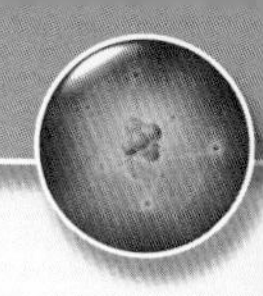

Eureka!

When an object is submerged in a fluid, it feels an upwards force caused by the fluid pressure – the **upthrust**. It turns out that the size of this force is equal to the weight of the fluid that has been displaced by the object. This is known as **Archimedes' Principle**. Thus, if the object is completely submerged, the mass of fluid displaced is equal to the volume of the object multiplied by the density of the fluid:

$$m = V\rho$$

The weight of fluid displaced (i.e. upthrust) is then found using the relationship:

$$W = mg$$

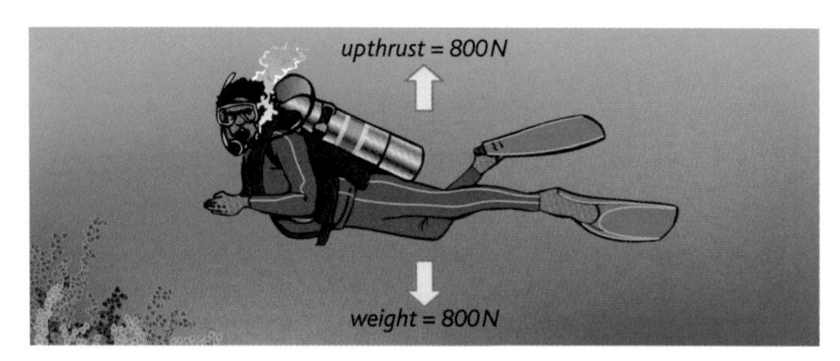

fig. 2.1.2 Scuba diving equipment includes a buoyancy control device which can change volume to displace more or less water. This varies the upthrust and so helps the diver move up or down.

This principle gets its name from the famous legend of the ancient Greek scientist Archimedes running naked from his bath through the streets of Syracuse in about 265 BC, shouting 'Eureka!' ('I've found it!'). According to this story, the king of Syracuse thought that his goldsmith may have stolen some gold, by creating a crown out of gold and silver mixed together and claiming it was pure gold. The king asked Archimedes to work out if this suspicion was true, without damaging the intricate wreath-style crown.

When Archimedes climbed into his bath, puzzling over the problem, the water overflowed. Observing this overflow, Archimedes realised that if the crown were submerged, it would displace its own volume of water and would experience an upwards force, or upthrust. Using a balance to weigh the crown when it was suspended in water, Archimedes could find the upthrust, and therefore the weight of water displaced and the volume of the crown. He could then calculate the density of the metal in the crown and compare it with the standard density for gold. He had solved the problem!

Why does a brick sink?

If the house brick from the example calculation of density on p. 52 were dropped in a pond, it would experience an upthrust equal to the weight of the volume of water displaced by the brick. As the density of water is $1000\,\text{kg}\,\text{m}^{-3}$, the mass of water displaced by the brick would be:

$$1000 \times 1.61 \times 10^{-3} = 1.61\,\text{kg}$$

This has a weight of:

$$1.61 \times 9.81 = 15.8\,\text{N}$$

so there is an upward force on the brick of 15.8 N.

If we compare the weight of the brick with the upthrust when it is submerged, the resultant force will be downwards:

weight $= 3.38 \times 9.81 = 33.2\,\text{N}$ downwards

upthrust $= 15.8\,\text{N}$ upwards

resultant force $= 33.2 - 15.8 = 17.6\,\text{N}$ downwards

So, the brick will accelerate downwards within the water until it reaches the bottom of the pond, which then exerts an extra upwards force to balance the weight so the brick rests on the bottom.

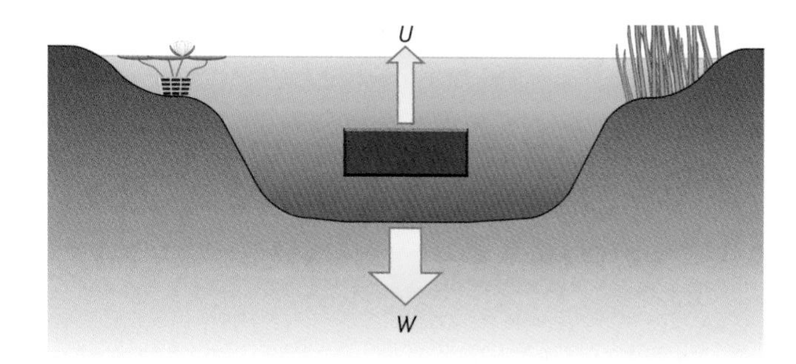

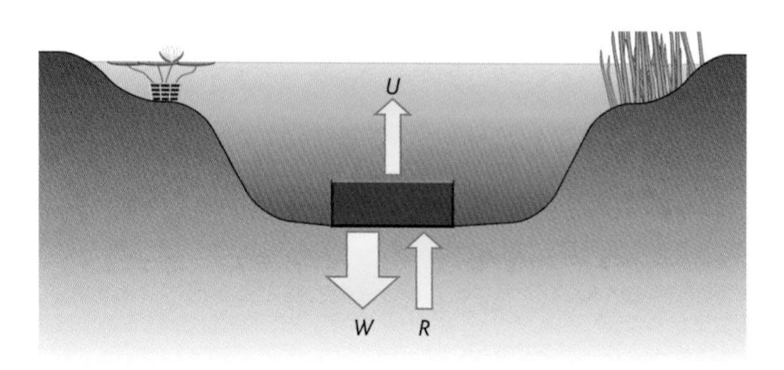

fig. 2.1.3 a If the upthrust on an object is less than its weight, then the object will sink through a fluid. b An object will remain at rest when balanced forces act on it.

Floating

Imagine an object falling into a fluid. When the object is at the surface there is no upthrust, because no fluid has been displaced. As the object sinks deeper into the fluid, it displaces a greater volume of the fluid, so increasing the upthrust acting upon it. When the upthrust and weight are balanced exactly, the object will float. So for an object to float, it will have to sink until it has displaced its own weight of fluid.

Worked examples

A giant garbage barge on New York's Hudson River is 60 m long and 10 m wide. What depth of the hull will be under water if it and its cargo have a combined mass of 1.5×10^6 kg? (Assume density of Hudson river water = $1000\,\text{kg m}^{-3}$.)

To float:

upthrust = weight

weight = $mg = 1.5 \times 10^6 \times 9.81 = 1.47 \times 10^7$ N

$\therefore$ upthrust = 1.47×10^7 N

The upthrust is equal to the weight of the volume of water displaced by the hull:

upthrust = $\rho \times V \times g$

where:

volume, V = length of hull, l × width of hull, w × depth of hull under water, d

So:

$$\text{upthrust} = 1000 \times 60 \times 10 \times d \times 9.81$$
$$= 5.89 \times 106 \times d$$
$$\therefore d = \frac{1.47 \times 10^7}{5.89 \times 10^6}$$
$$d = 2.497 \approx 2.5\,\text{m}$$

The hull will be 2.5 m under water.

HSW The Plimsoll Line

During the 1870s the MP Samuel Plimsoll fought a long struggle to pass a law in the English parliament to protect sailors. Merchant seaman had long known that overloaded ships were dangerous and prone to sinking in high seas. In the year 1873–1874 more than 400 ships were lost in the water around the United Kingdom, with a loss of over 500 lives. So-called 'coffin ships' became notorious. In some cases ships were so overloaded and in such a poor state of repair that sailors refused to put to sea in them and were imprisoned for desertion. Ship owners made huge profits from selling goods overseas, and if ships were lost at sea they would not lose money because they could claim on their insurance.

Plimsoll was determined to improve safety at sea. After many defeats, a law was passed in 1876 making it compulsory for every ship registered in England to be painted with a 'Plimsoll Line'. The mark shows the safe waterline on the hull of a loaded ship, and its correct position is worked out using density calculations .

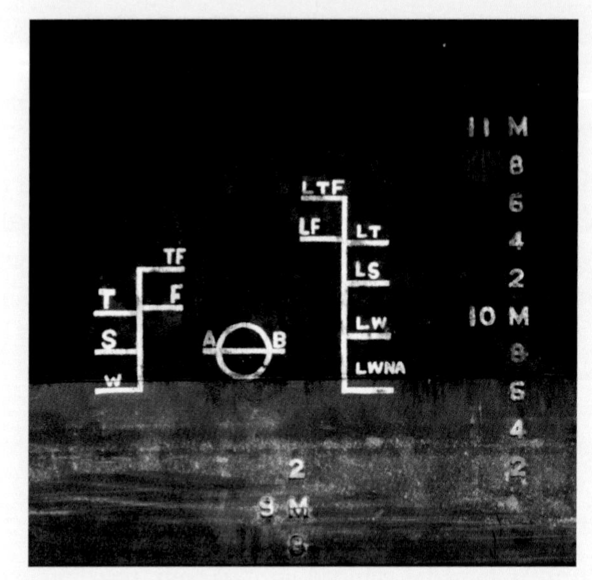

fig. 2.1.4 The Plimsoll Line on a ship indicates the safe loading level. The water surface must not be above the line indicating the temperature and location of the seas to be crossed (for example TF refers to tropical fresh water). As cargo is loaded, the ship will sink lower to displace a greater weight of water and thus balance the new, heavier overall weight of the ship.

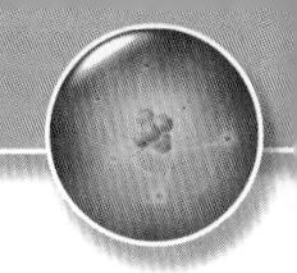

The hydrometer

The idea of floating at different depths is the principle behind the hydrometer, used to determine the density of a fluid. The device has a constant weight, so it will sink lower in fluids of lesser density. This is because a greater volume of a less dense fluid must be displaced to balance the weight of the hydrometer. Scale markings on the narrow stem of the hydrometer indicate the density of liquid.

Comparing the density of alcoholic drinks with that of water gives an indication of the proportion of alcohol they contain. The lower the density, the greater the alcohol content, as alcohol has a lower density than the water it is mixed with. This has long been the basis for the taxation of alcohol.

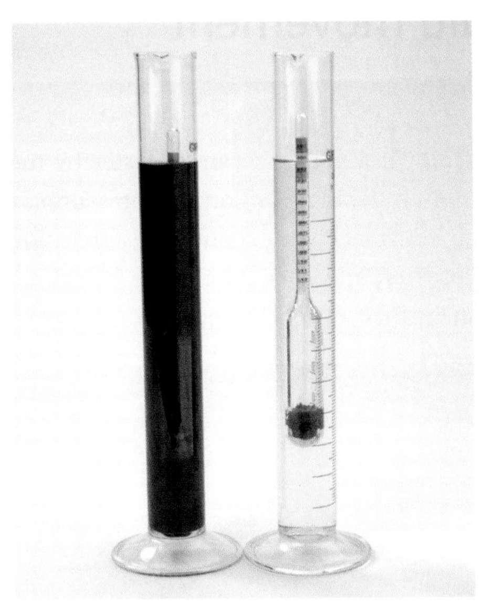

fig. 2.1.5 A hydrometer floats lower in the lower density (more alcoholic) red wine than in the white wine.

Type of drink	Alcohol content/abv (alcohol by volume, given as the percentage proportion of alcohol)	Taxation rate/£ per hectolitre (100 l) of product
Still cider and perry	1.2% < abv < 7.5%	26.48
Sparkling cider and perry	5.5% < abv < 8.5%	172.33
Wine	1.2% < abv < 4%	54.85
Sparkling wine	5.5% < abv < 8.5%	172.33

table 2.1.2 The greater the percentage of alcohol, the greater the tax on the drink.

Questions

1 A bottle of whiskey contains 1 litre of the drink. The mass of the liquid in the bottle is 0.915 kg. What is the density of this brand of whiskey? (1000 litres = 1 m^3)

2 The radius of a hockey ball is 36 mm and its mass is 140 g. What is its density
  - **a** in $g\,cm^{-3}$
  - **b** in $kg\,m^{-3}$?

3 Estimate the mass of air in this room. (Assume density of air = 1 $kg\,m^{-3}$.)

4 A golf ball has a diameter of 4.72 cm. If a golfer hits it into a stream, what upthrust does the ball experience when it is completely submerged? (Assume density of water = 1000 $kg\,m^{-3}$.)

5 Explain why a ship's Plimsoll Line has a mark for fresh water which is higher on the hull than the mark for salt water. (Assume density of salt water = 1100 $kg\,m^{-3}$.)

6 A ball bearing of mass 180 g is hung on a thread in oil of density 800 $kg\,m^{-3}$. Calculate the tension in the string, if the density of the ball bearing is 8000 $kg\,m^{-3}$.

2.2 Strength of materials

The physical properties of solids

Hooke's Law

There is a direct relationship between stretching a spring and the force it exerts. This relationship was first discovered by Robert Hooke in 1676 and is known as Hooke's Law. Hooke's Law states that the force F exerted by a spring is proportional to its extension Δx. This is usually written as:

$$F = -k\Delta x$$

where k is a constant for a particular spring called the constant of proportionality, or **spring constant**. The negative sign shows that the force exerted by the spring is in the opposite direction to the extension (fig. 2.2.1). By using the force exerted *by* the spring rather than the force exerted *on* the spring, we ensure that we are always focusing on two properties concerned with the spring itself, rather than one property of the spring (its extension) and a property of something else (the size of the load used to extend the spring).

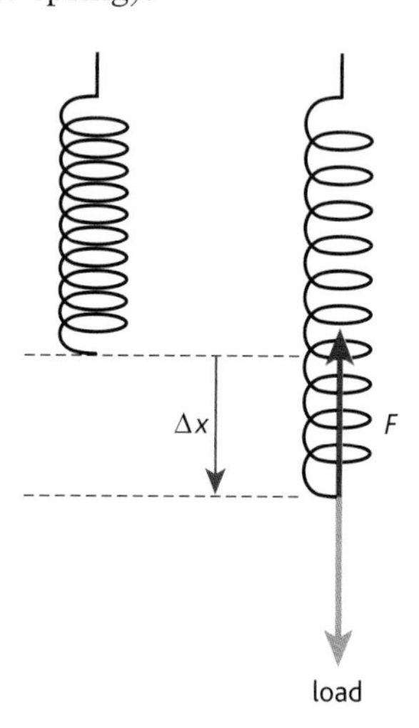

fig. 2.2.1 **Force and extension for a spring.**

Investigations show this law applies only for a load up to a certain limit. Up to this limit, the spring regains its original shape when the load is removed. Beyond this limit the extension increases more rapidly than expected, and the spring remains permanently deformed when the load is removed. The load at which this happens is called the **elastic limit**.

The spring constant k is different for different springs and different materials – the larger the value of k, the stiffer the spring (fig. 2.2.2). Although we do not usually make use of Hooke's Law when considering the stiffness of a particular material, solids do show very similar behaviour to springs. This provides evidence for a model of solids in which the attractive and repulsive forces between the particles behave a little like springs.

Beyond the elastic limit, materials no longer obey Hooke's Law and may be permanently deformed if stretched any further. This is known as **plastic deformation**. Some materials have a very low elastic limit and do not obey Hooke's Law at all. Plasticine is an example of such a material.

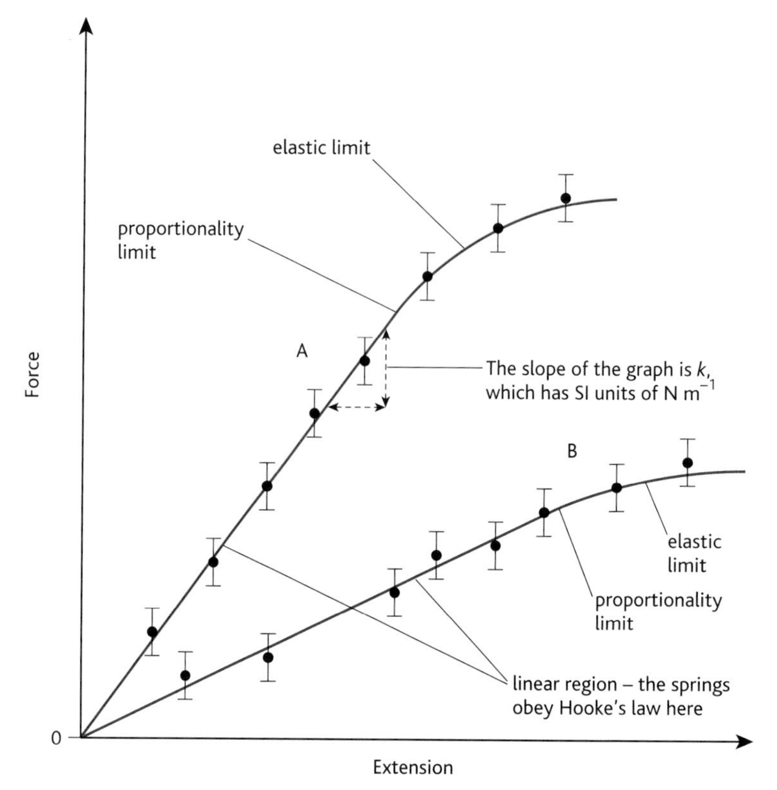

fig. 2.2.2 **Graphs of force against extension for two different springs. Spring A is stiffer than spring B – the curve has the greater slope, and so a greater value of k.**

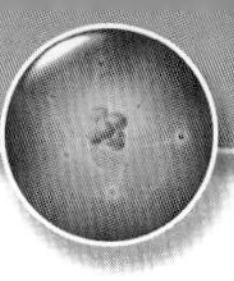

HSW Plotting graphs

Axes

In plotting a graph we represent the way in which one variable changes with respect to another variable. This is done by changing one variable (the **independent variable**) whilst keeping all other conditions constant, and making observations of the variable in which we are interested (the **dependent variable**). When these variables are plotted on a graph, the convention is to plot the **independent variable** on the *x*-axis, and the **dependent variable** on the *y*-axis. In the case of the force–extension graph for Hooke's Law, the dependent variable is the force exerted by the spring, which is measured for different extensions – hence this variable is plotted on the *y*-axis.

Note: In a school lab it is often easier to vary the force by loading a spring with an increasing number of slotted masses. In industrial materials testing labs, the extension is varied uniformly as the independent variable with the restoring force produced by the material being measured. So you may find that the graphs explained here seem 'backwards' compared with your own experiments!

Errors

Any investigation will involve uncertainty in measurements. These uncertainties arise in different ways (see HSW: Uncertainties in measurements on page 67). When plotting variables on a graph it is good practice to estimate the likely uncertainty in the measurement of the dependent variable, and to show this as a vertical error bar (see fig. 2.2.2). This can then be used when drawing a line through the points.

Lines

A line drawn through the points on a graph may be straight or curved, and it may pass through the origin or through some point on the *y*-axis. Before drawing a line through the points of a graph it is usually possible to have an intelligent idea about the likely shape and intercept of the line, particularly whether it should pass through the origin or not. Using the error bars, a line through the points can then be drawn, from which gradients and intercepts can be calculated.

Elastic strain energy

The energy involved in stretching a spring can be calculated if Hooke's Law is obeyed. For a given extension Δx, the force exerted by the spring is F. The energy stored in the stretched spring is equal to the work done on it as it is stretched – which is a result of the average force used to stretch it to the extension Δx. Since the force increases linearly from zero to F as the spring is stretched, the average force used to stretch it is given by:

$$\frac{0 + F}{2} = \tfrac{1}{2}F$$

So the work done on the spring is $\tfrac{1}{2}F\Delta x$. Since $F = -k\Delta x$, we can substitute for F in this relationship. This gives:

$$\text{work done} = \tfrac{1}{2}k\Delta x^2$$

This is expressed in terms of the extension of the spring and the force exerted *by* it – in other words, it is the work done *by the spring*. As this is *negative*, it means that the spring *gains* energy – so $\tfrac{1}{2}k\Delta x^2$ represents the energy stored in the spring – its potential energy. This potential energy is often called the **elastic strain energy**, E_{el}.

$$E_{el} = \tfrac{1}{2}F\Delta x = \tfrac{1}{2}k\Delta x^2$$

When we calculate the work done in stretching a spring, we are in effect calculating the area under the force–extension graph (fig. 2.2.3).

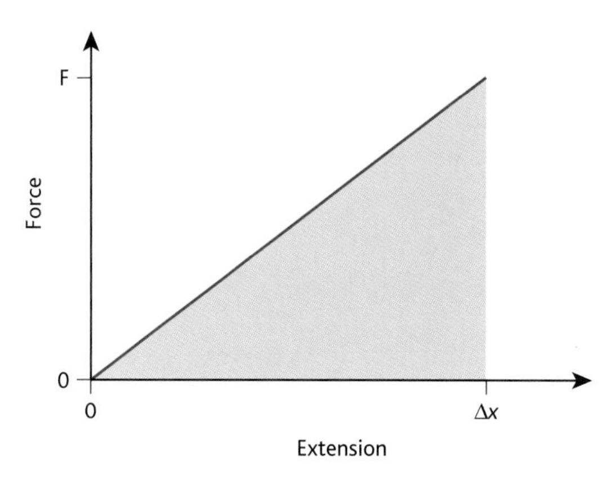

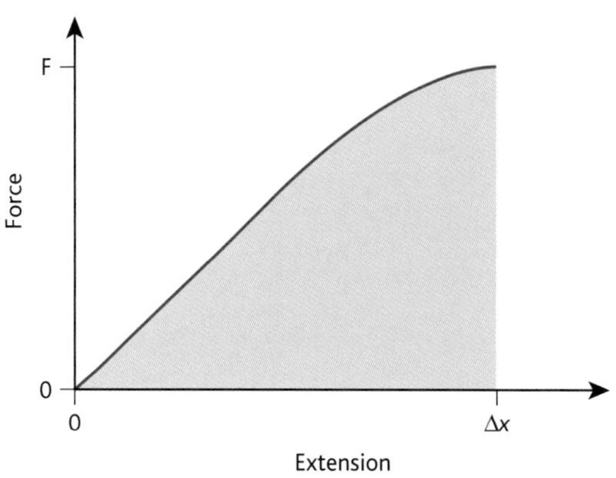

fig. 2.2.3 **The shaded area in each graph represents the work done in stretching a spring from zero extension to an extension of Δx.**

Worked examples

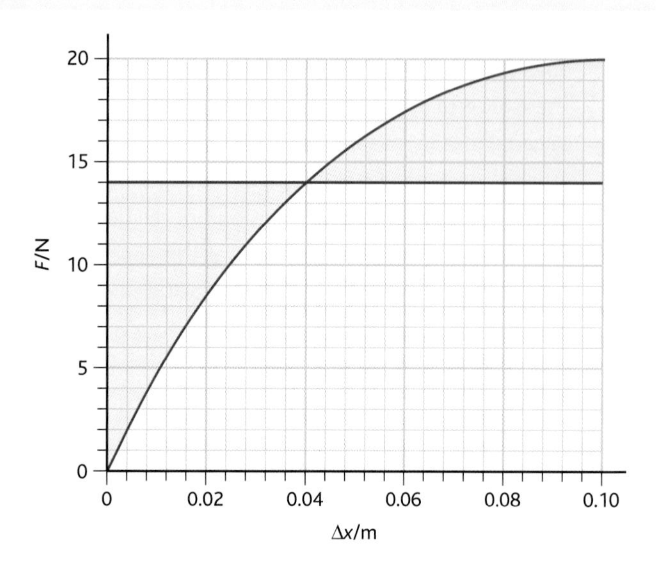

Fig. 2.2.4 shows a force–extension graph for a spring.

The force–extension graph is not a straight line, so we can't simply multiply the final force by the final extension to find the work done in stretching the spring. We must find the area under the graph by estimating.

One way to do this is to convert the shape into a rectangle. The blue line is drawn so that the two shaded areas above and below the curve are approximately equal. Then the area below the line represents the work done in stretching the spring. In **fig. 2.2.4** the blue line represents a force of 14 N, so the work done in stretching the spring is:

$$14 \times 0.1 = 1.4\,\text{N m} = 1.4\,\text{J}$$

This is the elastic strain energy stored in the spring.

Questions

1 In an experiment to find the spring constant for a certain spring, it was found to pull back with a force of 8 N, towards its original length of 18 cm, when stretched to a new length of 28 cm. Calculate the spring constant.

2 **Fig. 2.2.5** shows the result of an experiment in which the force exerted by a spring was measured for different extensions of the spring.

- **a** Explain how the graph shows that the spring obeys Hooke's Law.
- **b** Calculate the spring constant for this spring.
- **c** Calculate what the extension would be if the spring were exerting a force of 11.5 N
- **d** Calculate the elastic potential energy stored by the spring when it is extended by 5.5 mm.

3 An activity called 'Bungee Run' allows players to run along the ground whilst attached to an anchored bungee cord which has a spring constant of 25 N m^{-1}. If its natural length is 30 m and a runner manages to stretch it to a total length of 38 m, how much elastic potential energy is stored in the cord?

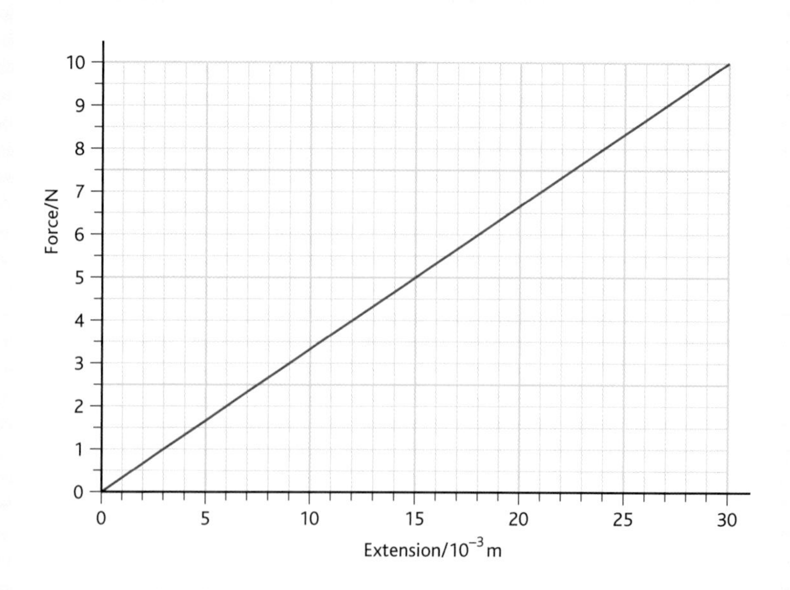

fig. 2.2.5 **Force–extension graph for a spring.**

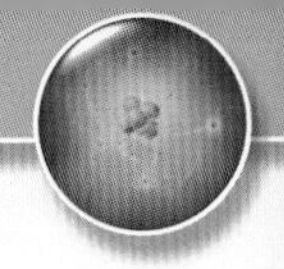

Stress, strain and the Young modulus

In stretching a spring we are applying a **tensile** force to it (a tensile force is one which puts something in **tension**, i.e. tends to pull it apart). A simple investigation shows that a metal wire behaves in a similar way to a spring when a tensile force is applied, initially obeying Hooke's Law.

fig. 2.2.6 The cables on a suspension bridge are under tension.

HSW Climbing ropes

If a manufacturer of climbing ropes is to select a material for a new rope they wish to make, what properties would they want the material to have? Apart from low cost, they will need to consider whether it will safely hold climbers, especially when they fall and the rope has to save them.

Knowing the amount of extension for a given force would be a useful start, and it could be helpful to know something about the force required to actually pull the material apart – its **tensile strength** (also called the **breaking stress**). Both of these will require us to know something about the size of the sample of material used, especially its area of cross-section, since this will have a great effect on its stretching behaviour and its strength. In a similar way, information about the length of the sample used is necessary if the information about the amount of stretch for a given force is to make any sense at all.

fig. 2.2.7 Ropes can save lives.

It can be easier to select the right materials if two new properties are defined so that the information about length and cross-sectional area can be included in the data relating to the samples.

- If we consider *tensile force per unit area* rather than tensile force alone, this takes into account the sample's area of cross-section.
- If we consider *extension per unit length* rather than extension alone, this takes into account the length of the sample.

Tensile force per unit area is called **tensile stress**, σ:

$$\text{tensile stress} = \frac{\text{tensile force}}{\text{area of cross-section}}$$

$$\sigma = \frac{F}{A}$$

The SI units of tensile stress are N m^{-2}, the same as those for pressure (so tensile stress is sometimes given in pascals). The **tensile strength** of a material is the tensile stress at which the material breaks.

Extension per unit length is called **tensile strain**, ε:

$$\text{tensile strain} = \frac{\text{extension}}{\text{original length}}$$

$$\varepsilon = \frac{\Delta x}{x}$$

Tensile strain has no units, as it is the ratio of two lengths.

Samples of many materials, particularly metals, are found to obey Hooke's Law for small tensile strains. Under these circumstances, the quantity:

$$\frac{\text{tensile stress}}{\text{tensile strain}}$$

is constant. This quantity is called the Young modulus, E.

$$E = \frac{\sigma}{\varepsilon}$$

The Young modulus has the same units as tensile stress, i.e. N m^{-2}. This is the same unit as pressure too, so you can quote both tensile stress and the Young modulus with the units of pressure – Pa.

Fig. 2.2.8 shows graphs of stress against strain for materials with different Young moduli. The **stiffer** a material, the greater its Young modulus. Table 2.2.1 lists the value of the Young modulus for different materials. The materials, A and B, shown on the graph in fig. 2.2.8 have only been tested here within their elastic limits. In the next section we will see how their behaviour changes, so that stress and strain are no longer proportional, when the elastic limit is exceeded.

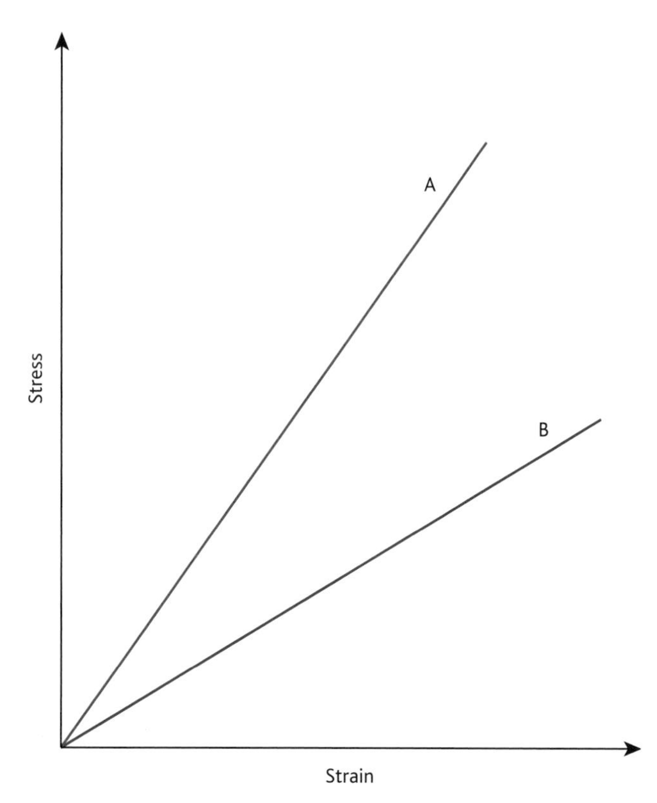

fig. 2.2.8 The Young modulus of a material is a measure of how stiff it is. Material A is stiffer than material B, as a higher stress is needed to produce the same strain.

Material	Young modulus /Pa	Example use
Mild steel	2×10^{11}	Cables
Aluminium	7×10^{10}	Aircraft
Softwood (parallel to grain)	1.6×10^{10}	Construction
Brick	7×10^{9}	Construction
Concrete	4×10^{10}	Construction

table 2.2.1 The Young modulus of a material is one piece of information used by engineers in selecting materials for particular uses.

PB Finding the Young modulus by experiment

Stress–strain curves can be produced from simple experiments in a school laboratory and also allow us to find the Young modulus for common materials like copper. Hanging weights can be used to apply a tension force on a wire. Careful measurement of the extension can be used with measurements of the original length to find the strain. A micrometer screw gauge will measure the wire diameter, from which you can calculate the cross-sectional area.

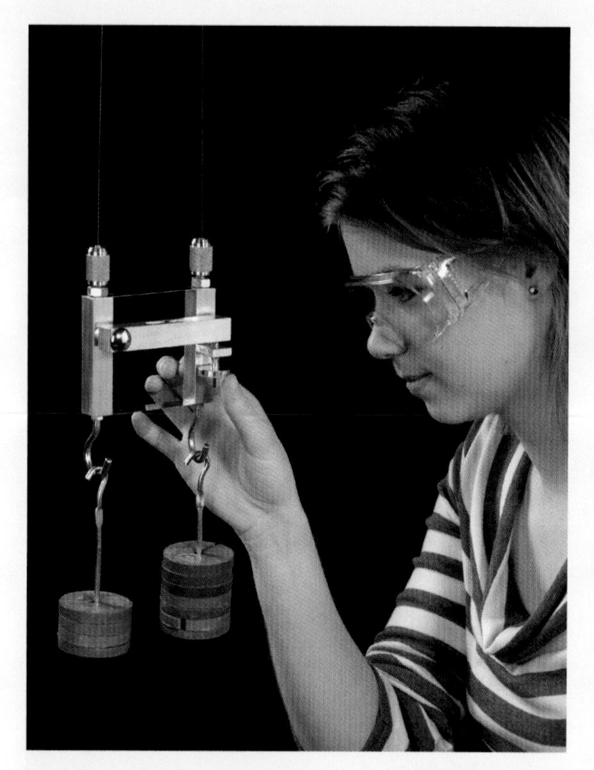

fig. 2.2.9 Searle's apparatus can be used to find the Young modulus.

By plotting the stress–strain curves from experiments like these, you can investigate the elastic and plastic deformation of a material, and distinguish between them.

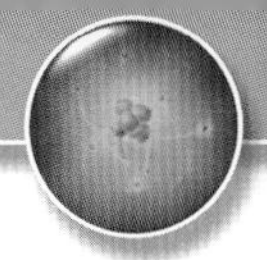

HSW Uncertainties in measurements

To measure the Young modulus all that is necessary in principle is to measure the extension of a sample for a given force and then to calculate the tensile stress and tensile strain, from which E can be calculated. In practice such a method is unlikely to lead to very reliable results.

To find E, a **graphical** method is usually used. Since:

$$E = \text{tensile strain} = \frac{\sigma}{\varepsilon}$$

then

$$\sigma = E\varepsilon$$

It follows that a graph of tensile stress plotted against tensile strain will have slope E (compare $\sigma = E\varepsilon$ with $y = mx + c$).

To plot the graph you will need to make a number of measurements. Table 2.2.2 suggests possible **uncertainties** in these measurements when using simple equipment.

Measure-ment	Possible uncertainty in measurement	Note
Force/N	±2% (if using slotted masses)	Slotted masses may have very variable uncertainties in their mass. If carefully selected, they are unlikely to have an uncertainty of more than ±2 g. The mass used will need to be converted to weight using an appropriate value of g – the approximation of g to 10 N kg^{-1} would lead to an uncertainty of more than +1%.
Diameter of wire/m	±2% (using micrometer)	A micrometer should be capable of measuring a wire 1 mm in diameter to within ±0.02 mm if used carefully. This is an uncertainty of ±2%, which will also be the uncertainty of the radius (d/2).
Cross-sectional area/m^2	±4% (depends on radius^2)	Since the cross-sectional area of the wire depends on its radius^2, the uncertainty of the cross-sectional area will be *twice* the uncertainty of the radius. (Where two quantities are multiplied together or one is divided by the other their uncertainties are *added* to find the uncertainty of the result. Where two quantities area added or subtracted their uncertainties are *averaged* to find the uncertainty of the result.)
Original length of wire/m	±1.5% (using metre rule)	If a piece of wire 2 m long is used, its length can be measured to within around 3 mm using a metre rule – an uncertainty of 1.5%.
Extension of wire/m	±10%	The extension of the wire depends on many factors like its length, cross-sectional area and the material it is made of. If extensions of the order of 5 mm are measured to within 0.5 mm, the uncertainty is ±10%.
Tensile stress = F/A	±6%	This is the sum of the uncertainties for F and A.
Tensile strain = $\Delta x/x$	±11.5%	This is the sum of the uncertainties for Δx and x.

table 2.2.2 **Measurements for finding E, together with possible uncertainties.**

The largest uncertainty in any measurement of E comes in the measurement of extension. Accurate methods for measuring E thus concentrate on measuring Δx with as great a precision as possible. The percentage errors are calculated by finding what fraction of the value the error represents. For example, for the error in the extension of the wire:

error in measurement = ± 0.5 mm

value of measurement ≈ 5 mm

$$\%\ \text{error} = \frac{0.5}{5} \times 100\%$$

$$= 10\%$$

The overall uncertainty in E is likely to be ±6% + ±11.5%, i.e. ±17.5%.

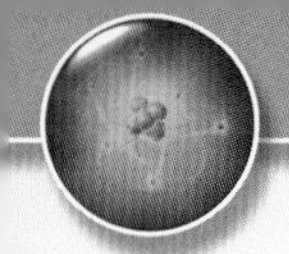

Characteristics of solids

The behaviour of materials which are subject to stress is often represented by means of stress–strain curves. Fig. 2.2.10 shows a typical stress–strain curve for a metal. As the stress increases, the sample begins to narrow at one point. This is called 'necking' (fig. 2.2.11).

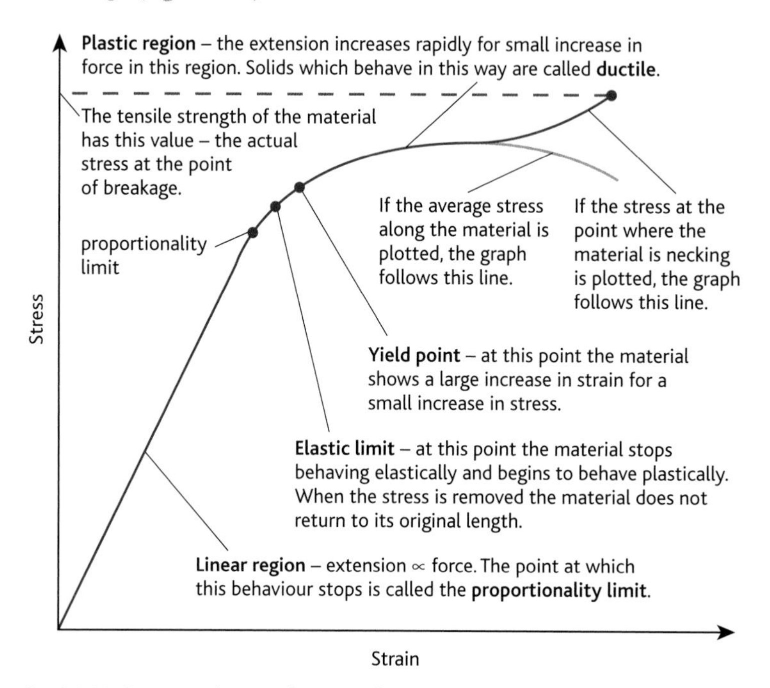

fig. 2.2.10 Stress–strain curve for a metal.

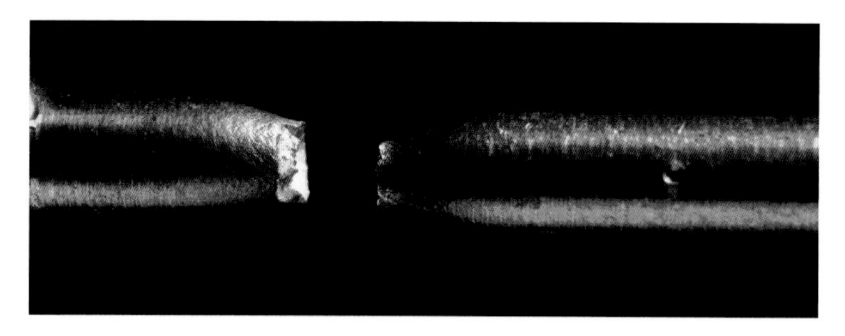

fig. 2.2.11 Necking in a wire under stress. Because the stress is proportional to 1/area, the stress is larger at this point than at other points along the sample. Once necking has occurred, the sample will begin to fail at this point.

Deforming solids

So far we have looked at the ability of a material to resist a tensile force, which is called its **stiffness**, and the tensile stress at which a material fails – its **tensile strength**. Both of these terms consider only situations in which the sample of material is under tension (i.e. being pulled apart), without thinking about any other situation. In many situations, the force on a material will be tending to reduce the volume – to squash the material. This is known as a **compressive** force and puts the material under **compression**.

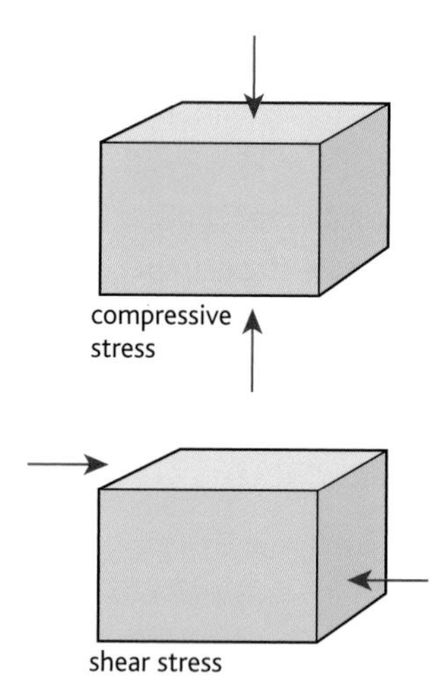

fig. 2.2.12 In practical situations, materials need to be able to resist these sorts of deformations as well as tensile deformations.

Compressive force per unit area is called **compressive stress**:

$$\text{compressive stress} = \frac{\text{compressive force}}{\text{area of cross-section}}$$

The SI units of compressive stress are N m^{-2}, the same as those for pressure (so compressive stress is sometimes given in pascals).

The **compressive strength** of a material is the compressive stress at which the material breaks.

Extension per unit length is called **compressive strain**:

$$\text{compressive strain} = \frac{\text{extension}}{\text{original length}}$$

Compressive strain has no units, as it is the ratio of two lengths.

Some materials have very low tensile strength, but are strong when they are subjected to compressive stress – brick and concrete are two very common examples of such materials. The strength of a material under **shear stress** is related to some extent to its tensile strength. In general, the **strength** of a material refers to its ability to withstand stress, whether tensile, compressive or shear.

Worked examples

A lift is designed to hold a maximum of 12 people. The lift cage has a mass of 500 kg, and the distance from the top floor of the building to the ground floor is 50 m.

a What minimum cross-sectional area should the cable have?

b Estimate how much the steel lift cable will stretch if 10 people get into the lift at the ground floor, assuming that the lift cable has four times the area of cross-section calculated in **a**.

(Young modulus of steel = $2.0 \times 10^{11}\,\text{N m}^{-2}$, tensile strength of steel = $4.0 \times 10^{8}\,\text{N m}^{-2}$, $g = 9.81\,\text{N kg}^{-1}$.)

a Assume the mass of an average person = 70 kg

load on cable = $[(12 \times 70\,\text{kg}) + 500\,\text{kg}] \times 9.81\,\text{N kg}^{-1}$

$= 13\,000\,\text{N}$

tensile strength = $4.0 \times 10^{8}\,\text{N m}^{-2}$

$= \dfrac{13\,000\,\text{N}}{\text{area}}$

Rearranging:

$$\text{area} = \frac{13\,000\,\text{N}}{4.0 \times 10^{8}\,\text{N m}^{-2}} = 3.3 \times 10^{-5}\,\text{m}^2$$

Minimum area of cross-section of cable is $0.33\,\text{cm}^2$.

b Assume the mass of an average person = 70 kg.

extra load on cable = $10 \times 70\,\text{kg} \times 9.81\,\text{N kg}^{-1}$

$= 6900\,\text{N}$

cross-sectional area of cable = $4 \times 3.3 \times 10^{-5}$

$= 1.3 \times 10^{-4}\,\text{m}^2$

$$E = \frac{\text{tensile stress}}{\text{tensile strain}} = \frac{\text{tensile force/area}}{\text{extension/original length}}$$

so

$$2 \times 10^{11}\,\text{N m}^{-2} = \frac{6900\,\text{N}/1.3 \times 10^{-4}\,\text{m}^2}{\Delta x/50\,\text{m}}$$

Rearranging:

$$\Delta x = \frac{6900/1.3 \times 10^{-4}}{2 \times 10^{11}/50}$$

$= 1.3 \times 10^{-2}\,\text{m}$

The lift cable stretches by about an additional 1 cm. (Note that the lift cage will already have stretched the cable by a little under 1 cm.)

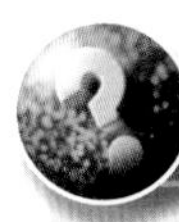

Questions

1 Explain why the units of the Young modulus could be Pascals.

2 The wire in a new guitar 'string' is made of a 90 cm steel wire of diameter 1 mm. When the string is fitted to the guitar, the string is put under a tension of 75 N by winding it round a peg and this also stretches it by 0.5 mm.

a What is the stress in the wire?

b What is the wire's strain?

c What is the Young modulus for this kind of steel?

3 a 'In an experiment to find the Young modulus, the strain should not be more than 1 in 1000'. Explain what this statement means.

b Describe an experiment to determine the Young modulus for the metal in a wire. Taking into account the possible errors in measurements, explain why the limitation in **a** is necessary.

c In such an experiment, a brass wire of diameter $9.50 \times 10^{-4}\,\text{m}$ is used. If the Young modulus for brass is $9.86 \times 10^{10}\,\text{N m}^{-2}$, find the greatest force which could be used to keep within the limitation in part **a**.

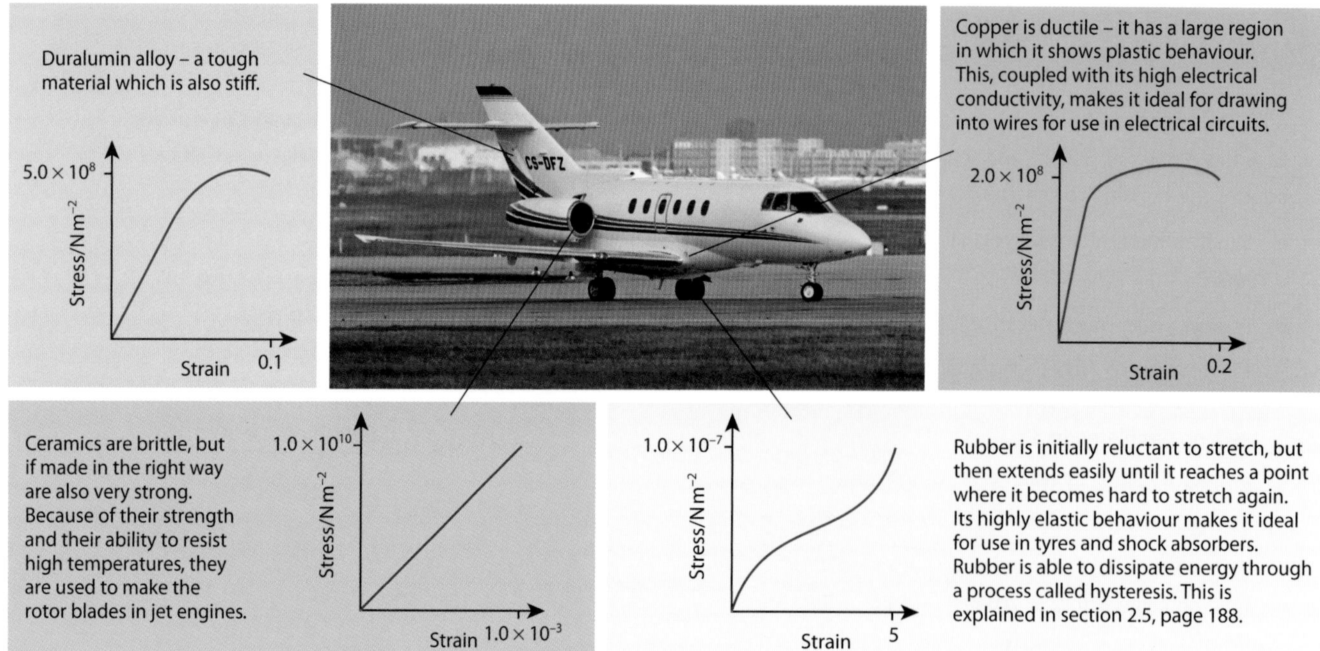

fig. 2.2.13 The usefulness of a material depends on its stress–strain behaviour.

Ductile materials are those which show plastic deformation – metals usually behave like this (fig. 2.2.10). Materials which break or crack with little deformation are called **brittle**. It is important to understand that brittle does not mean **weak** – although glass is a brittle material, its tensile strength can be as great as that of aluminium. Some ceramics (materials generally made of clay) actually have a greater tensile strength than most metals!

Tough materials are able to withstand impact forces without breaking and require a large force to produce a small plastic deformation. Tough materials are therefore not brittle. Composite materials, which are combinations of more than one material, often gain the best properties of both. Living wood is an example of a composite material which is tough – the combination of materials within the wood allow it to be both strong and elastic, which makes it very tough. A lot of energy is required to break wood from a live tree. Once dead, wood dries out and this removes the flexibility, making it more brittle – and less tough. Sports equipment is often made from carbon fibre (another composite material), as this is also tough and will survive the rigours of the sport without breaking.

Hard materials are those which resist plastic deformation – usually by denting, but you can also consider it in terms of scratching or cutting. There are various standard techniques for measuring this, most of which involve measuring the size of a dent produced by pressing a diamond into the surface with a certain force.

In 1824, the German geologist Friedrich Mohs published a scale for classifying mineral hardness. His principle was that a material that could scratch another material should be higher (or at least the same value) on the scale. He standardised the scale by naming ten types of mineral in order, from talc which is the softest mineral, up to diamond, which is the hardest natural material. It is for this reason that diamond is used in some hardness testing procedures, and is commonly used for industrial drill bits.

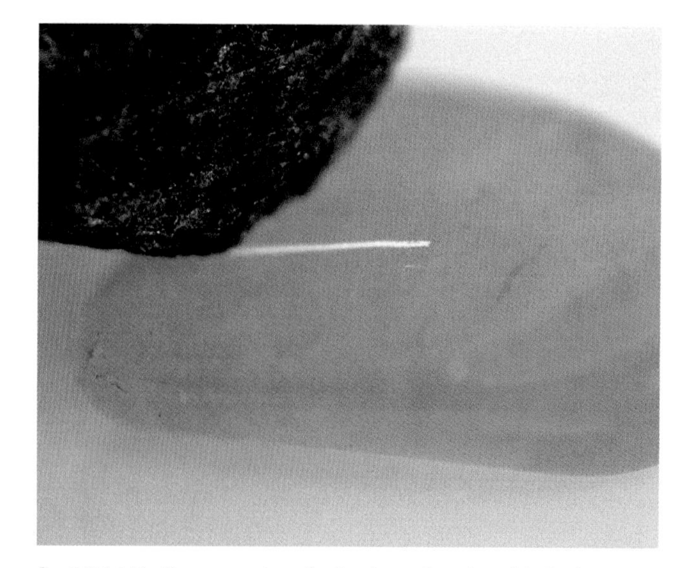

fig. 2.2.14 Testing a gemstone for hardness. Starting with the lowest hardness value, sharp instruments of various hardnesses are applied to the lowest part of the gem until a scratch is made.

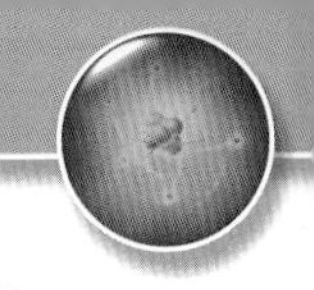

The scale is effective but, by comparing materials with each other, does not give absolute values and so is little used in engineering.

Mohs hardness number	Mineral
1	Talc
2	Gypsum
3	Calcite
4	Fluorite
5	Apatite
6	Orthoclase
7	Quartz
8	Topaz
9	Corundum
10	Diamond

table 2.2.3 Harder materials can scratch softer ones. For example, if a test material can scratch calcite but is itself scratched by fluorite, then it would be given a value of 3.5 on the Mohs hardness scale.

Materials which show large plastic deformation before cracking or breaking are called **malleable**. These are therefore not brittle, although they are not necessarily tough. It is a similar property to ductility, as a malleable material can be reshaped easily without fracturing. The word comes from *malleus*, the Latin word for hammer, as malleable materials can be worked into shape with a hammer. The most malleable material is gold, which can be rolled, pressed or hammered with the greatest of ease. Sheets as thin as 0.2μm, known as gold leaf, are used artistically in gilding. It can also be so thin that the quantity of metal is so small that it is considered safe for human consumption and is used to decorate chocolates and desserts, or even drinks. Also, the minute quantity of gold used means that these uses may not be prohibitively expensive.

HSW The Mohs hardness scale

There is evidence that Mohs took considerable credit for work that was not entirely his own. A system for comparing the hardness of minerals very similar to Mohs' was published by Abraham Werner in 1774, when Werner was 24 years old. Werner went on to become a professor at Freiburg University, and Mohs' teacher. This system of classifying mineral hardness did not become widely used until Mohs, who had stepped into Werner's job after his death in 1817, published a very similar idea in 1824.

HSW Materials selection charts

Scientists will often come up with a need for a material with certain properties. In order to make a good climbing rope, the manufacturers will want a material which, amongst other properties, is lightweight, and stretches a little but not too much – a medium Young modulus. In order to choose an appropriate material to solve a problem, they may look at a materials selection chart. This has two properties plotted as graph axes and various materials are positioned on the chart depending upon their values for those two properties.

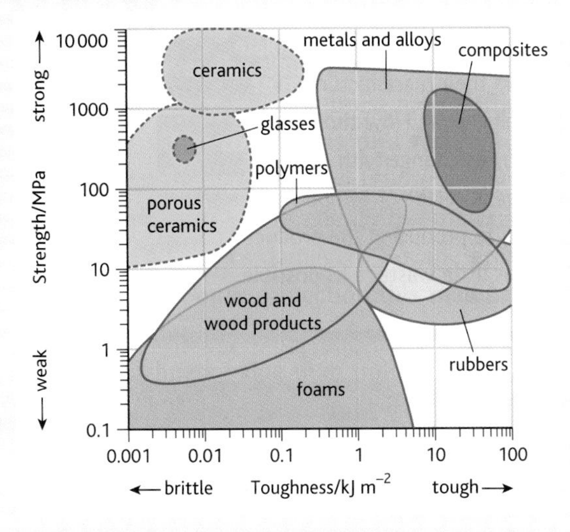

fig. 2.2.15 A chart of strength against toughness is useful for selecting materials that are not only strong (can carry a high static load before deforming) but also tough (can withstand impact loads).

Questions

1 What is meant by **a** elastic limit
b plastic behaviour **c** Hooke's Law
d breaking stress **e** compressive strain?
Illustrate your answers with appropriate graphs.

2 The compressive strength for marble is 112 MPa. If a marble column holding up the roof of a temple is a cylinder and supports 1.43×10^7 N of the weight of the roof, what is its minimum diameter?

3 Define the following terms and, for each, give an example of a material which shows the property and a use that this makes it suitable for:
a malleable **b** tough **c** hard **d** ductile
e brittle

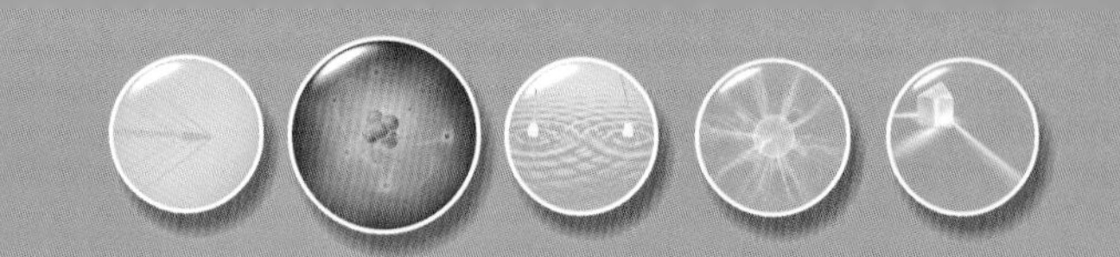

Examzone: Topic 2 Materials

1 Control of high-volume manufacturing production, such as in the steel industry, is achieved through regular sampling and testing of the product.

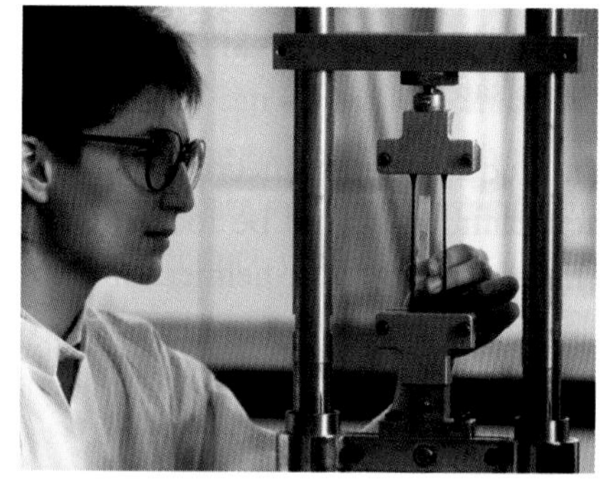

The picture shows a machine called a tensile tester. It is stretching a sample at a constant rate. The test sample is a rod of steel approximately the size of a pencil.

The results below were from a test on a sample of steel of 1.3×10^{-4} m^2 cross-sectional area and 6.5×10^{-2} m length. The tension T applied to the sample and its resulting extension Δx were measured until the sample failed.

$T/10^3$ N	0	5	10	15	20	25	30	35
$\Delta x/10^{-6}$ m	0	12	24	36	48	60	74	100

Plot a graph of these results. **(3)**

Indicate on the graph with the letter P the limit of proportionality. **(1)**

Calculate the stress applied to the specimen at this point. **(2)**

Calculate the strain in the sample at point P. **(1)**

Calculate the Young modulus for this steel. **(1)**

A second sample of exactly the same size is stiffer, weaker and brittle.

Sketch a line on your graph predicting the results for the sample. Label this line X. **(2)**

(Total 10 marks)

2 Nomlas is a new material intended for fishing rods.

Copy and complete the table below for the four other properties of materials listed.

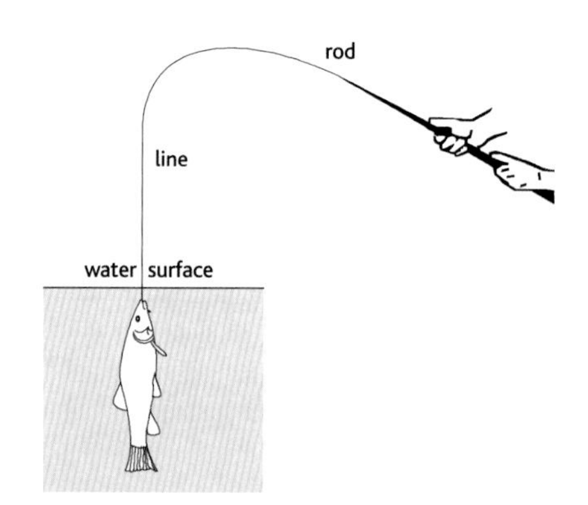

Property	Desirable for rod	Not desirable for rod	Reason
Strong	✓		Needs a large force before it breaks
Elastic			
Brittle			
Hard			
Tough			

(Total 8 marks)

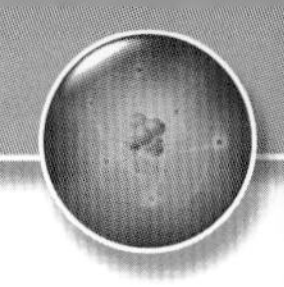

3 Training shoes (trainers) have changed a lot from the original rubber-soled canvas shoes of the 1930s. They now combine the most up-to-date research from the fields of physics and chemistry to cope with high levels of impact.

Copy and complete the following table for the material properties listed.

Property	Desirable for trainers	Not desirable for trainers	Reason
Stiff		✓	Needs a large force to produce a small deformation
Plastic			
Tough			
Brittle			

(6)

Calculate the average compressive stress exerted on the soles of the trainers worn by an 80 kg athlete when the athlete is standing still.

Total area of both trainer soles in contact with the ground = 4.2×10^{-2} m^2. **(2)**

At certain positions during the athlete's running stride the stress is greater than when the athlete is standing still. Identify one such position and explain why the stress is greater. **(2)**

(Total 10 marks)

4 Do not try this at home!

The website 'urban myths' claims that a man in California tied a number of balloons filled with helium to his chair in the garden, with a view to gently hovering above the neighbourhood.

The moment he cut the anchoring cord he shot upwards to a height of about 4000 m. Several hours later he was rescued by a helicopter after being spotted by an airline pilot.

If the combined mass of the man and the chair was 70 kg, calculate their weight. **(1)**

What is meant by the term **upthrust**? **(2)**

Show that the upthrust in newtons from the balloons is about $1.3V$ where V is the total volume of the balloons in cubic metres. The density of air is 1.29 kg m^{-3}. **(2)**

Write down an expression, in terms of V, for the weight of the helium in the balloons. The density of helium is 0.18 kg m^{-3}. **(1)**

Calculate the total volume of the balloons required just to lift the man and his chair from the ground. Assume the weight of the balloon fabric is negligible. **(3)**

Explain why any viscous drag force can be ignored in the previous calculation. **(2)**

(Total 11 marks)

Topic 3 Waves

This topic covers the properties of different types of waves, including standing (stationary) waves. You will also explore applications of waves in music, medical physics and astronomy.

What are the theories?

Before you can make any detailed interpretation of waves, you will need to learn the fundamentals of how to describe them and measure their properties, such as wavelength, frequency, speed and amplitude. These help us to classify waves – in particular the various regions in the electromagnetic spectrum.

There are many interesting wave phenomena. How effects such as diffraction and polarisation can be explained will be covered in some detail. You will learn how to calculate the degree to which waves are refracted using Snell's law, and how observations of the Doppler effect provide evidence for the Big Bang theory and the expanding Universe.

What is the evidence?

You will investigate various aspects of waves, including experiments to confirm the refractive index of materials, and predict whether total internal reflection will occur. You will also undertake investigations into diffraction and polarisation.

Scientists have used experimental evidence about waves to confirm many theories about the world around us, and this has led to refinement of the details of our understanding of the nature of the Universe. For example, our explanation of the nature of the electron has been updated after scientists managed to diffract electrons.

What are the implications?

The differing properties of the different regions of the electromagnetic spectrum make them useful for differing applications. You will get to grips with some applications for each part of this spectrum. For example, the total internal reflection of visible light allows it to travel along an optical fibre, carrying information for telecommunications, or as a picture of a place your eye could not reach, like the inside of a patient or the bottom of a narrow drain.

The wave explanation of how bats navigate has led scientists to the development of medical ultrasound diagnosis techniques. These are now so sophisticated that pregnant mothers can obtain an ultrasound 'video' of their unborn child.

Edwin Hubble's analysis of the Doppler shift of light from distant galaxies led to the development of the Big Bang theory and estimations of the age of our Universe.

The map opposite shows you all the knowledge and skills you need to have by the end of this topic. The colour in each box shows which chapter they are covered in and the numbers refer to the sections in the Edexcel specification.

Chapter 3.1

use graphs to represent transverse and longitudinal waves, including standing waves (32)

explain and use the concepts of wavefront, coherence, path difference, superposition and phase (33)

recognise and use the relationship between phase difference and path difference (34)

recall that a sound wave is a longitudinal wave which can be described in terms of displacement of molecules (31)

understand and use the terms *amplitude*, *frequency*, *period*, *speed* and *wavelength* (28)

use the wave equation $v = f\lambda$ (30)

explain what is meant by a *standing* (*stationary*) wave, investigate how such a wave is formed, and identify nodes and antinodes (35)

identify the different regions of the electromagnetic spectrum and describe some of their applications (29)

Chapter 3.2

recognise and use the expression for refractive index ${}_1\mu_2 = \sin i/\sin r = v_1/v_2$, determine the refractive index for a material in the laboratory, and predict whether total internal reflection will occur at an interface using critical angle (36)

investigate and explain how to measure refractive index (37)

discuss situations that require the accurate determination of refractive index (38)

investigate and recall that waves can be diffracted and that substantial diffraction occurs when the size of the gap or obstacle is similar to the wavelength of the wave (41)

explain how diffraction experiments provide evidence for the wave nature of electrons (42)

recall that, in general, waves are transmitted and reflected at an interface between media (44)

investigate and explain what is meant by *plane polarised light* (39)

discuss how scientific ideas may change over time, for example our ideas on the particle/wave nature of electrons (43) (also Chapter 3.3 – light as an electromagnetic wave and the discovery of the Hubble red shift as evidence for an expanding Universe)

explain how different media affect the transmission/reflection of waves travelling from one medium to another (45)

investigate and explain how to measure the rotation of the plane of polarisation (40)

Chapter 3.3

identify the different regions of the electromagnetic spectrum and describe some of their applications (29)

explore and explain how a pulse-echo technique can provide details of the position and/or speed of an object and describe applications of this technique (46)

explain how the amount of detail in a scan may be limited by the wavelength of the radiation or by the duration of pulses (48)

use the wave equation $v = f\lambda$ (30)

explain qualitatively how the movement of a source of sound or light relative to the observer/detector gives rise to a shift in frequency (Doppler effect) and explore applications that use this effect (47)

discuss the social and ethical issues that need to be considered, for example when developing and trialling new medical techniques on patients (49)

3.1 The language of waves

Types of wave

All waves consist of some sort of disturbance travelling through space. Some waves require a material substance (called a **medium**) through which to travel. Waves like this are called **mechanical waves** – examples of these include water waves, sound waves and the waves that travel through the Earth's crust during earthquakes (**seismic waves**). **Electromagnetic waves**, like infrared for example, require no medium through which to travel – the heat from the Sun reaches us through the vacuum of space. We shall examine both of these types of wave during the course of this topic.

If we wish to classify waves into different types, one starting point is to draw the distinction between **transverse** and **longitudinal** waves. This classification is based on the way that the two types of wave travel.

All waves happen through some sort of disturbance. For all waves, something oscillates in order to allow the transfer of energy. In a transverse wave, the direction in which disturbance takes place is at right angles to the direction in which the wave travels. (The direction of travel is referred to as the direction of **propagation** of the wave). In a longitudinal wave, the disturbances take place in a direction parallel to the direction of propagation of the wave. Both types of wave can be demonstrated using a long steel spring called a 'slinky'. This is shown in **fig. 3.1.1**.

To make a transverse wave travel along the slinky, the end of the spring is moved to and fro at right angles to the length of the spring.

To make a longitudinal wave travel along the slinky, the end of the spring is moved to and fro parallel to the length of the spring.

In a transverse wave, the oscillations occur at right angles to the direction of travel of the wave.

In a longitudinal wave, the oscillations occur parallel to the direction of travel of the wave. As a result, a longitudinal wave consists of a series of **compressions** (C) and **rarefactions** (R).

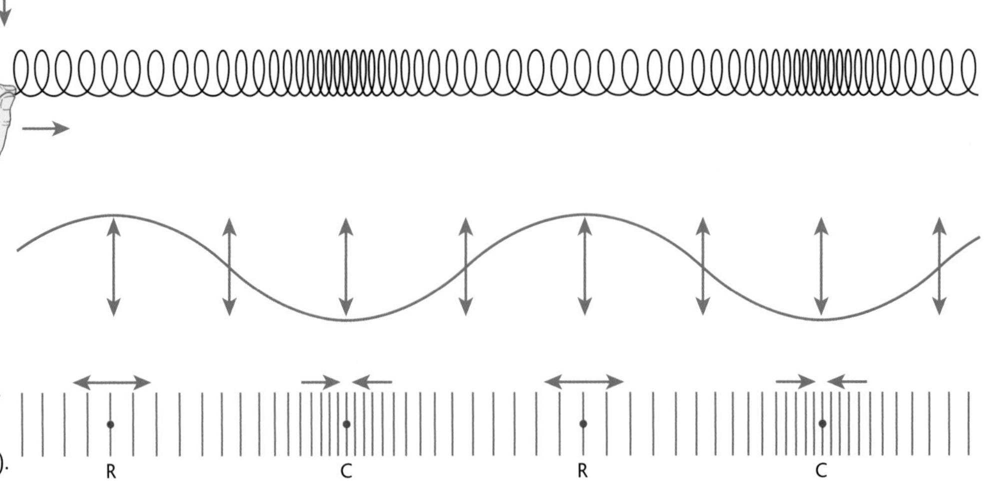

fig. 3.1.1 The propagation of a transverse wave and a longitudinal wave along a slinky. In both cases the wave carries energy from left to right, away from the source of the waves.

Waves that are produced on a slinky in this way travel from one end of it to the other – they are examples of **progressive** waves.

Waves – circular and plane, continuous, trains and pulse

The way in which a wave is described in physics gives some very precise information about it. Seen from above, the waves that result from a stone being dropped into a pond appear as a series of circles, like the waves on the ripple tank in **fig. 3.1.2a**. Waves like this are called **circular waves**. Waves on the surface of the sea often appear as a series of parallel lines, like those on the ripple tank in **fig. 3.1.2b**. These are **plane waves**.

The waves on the surface of the sea may seem to go on for ever, but of course they do not! Physicists often want to refer to waves that *do* go on for ever, i.e. that have an infinite length. Such waves are the products of your imagination, but are often used in situations where physicists want to be sure that they are modelling what is happening in the *middle* of a wave rather than at the start or end of it. Waves like this are called **continuous waves**, while waves that *do* have a beginning and an end are called **wave trains**. Where we are considering a very short wave motion, the term **pulse** is used – a wave pulse does not contain any repeated up and down motion.

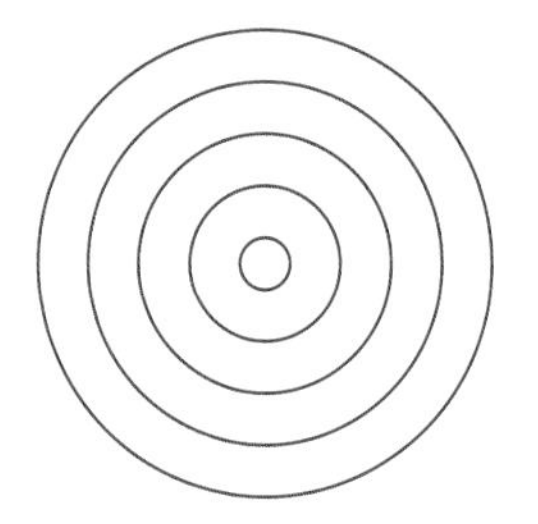

fig. 3.1.2a Circular waves on a ripple tank. In the diagram representing the waves, the lines represent the crests of the waves, and are referred to as **wavefronts**. b Plane waves on a ripple tank and their diagrammatic representation. Plane waves can be thought of as a special case of circular waves – they are effectively circular waves that have their source a very long distance away.

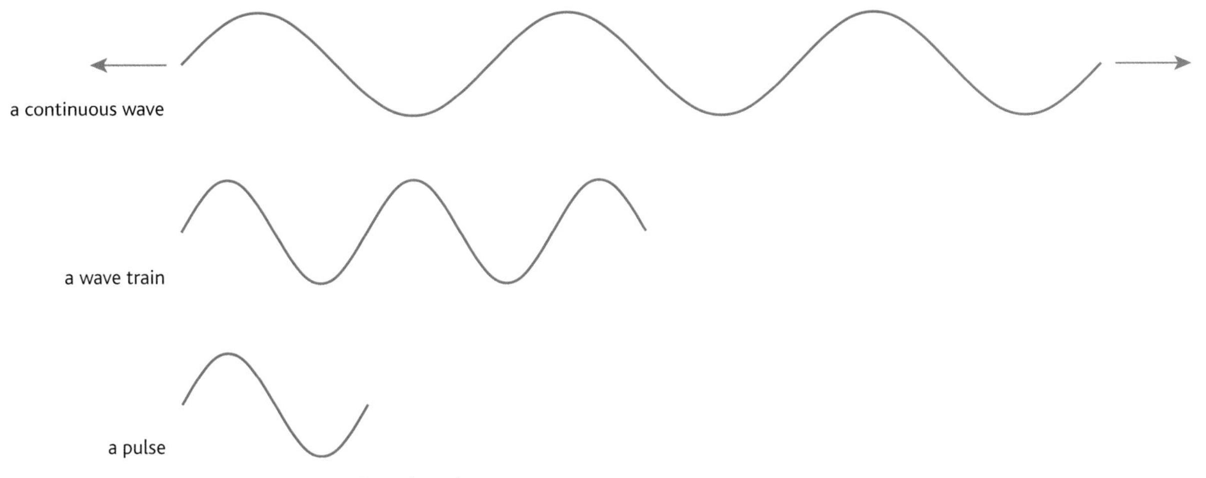

fig. 3.1.3 A continuous wave, a wave train and a pulse.

An introduction to the behaviour of waves

The principle of superposition

fig. 3.1.12 Waves can cross each other's paths without disturbing each other.

When two wave pulses are sent from opposite ends of a slinky spring, the pulses travel through each other and on to the end of the spring as if the other pulse was not there at all. If both pulses have the same phase, at the point where the two cross a large pulse can be seen for a brief instant, before the two pulses continue on their way. On the other hand, if the two pulses are 180° out of phase the spring appears undisplaced while the two pulses cross (**fig. 3.1.13**).

This behaviour of waves can also be observed using water waves on a ripple tank, although it is not quite so easy to do. The behaviour is summarised as the **principle of superposition**, which states that:

> **Where two or more waves meet, the total displacement at any point is the sum of the displacements that each individual wave would cause at that point.**

(Since displacement is a vector quantity, in determining the total displacement it is important to remember to take into account whether each individual displacement is positive or negative.)

The principle of superposition applies to *all* waves, and depends on the phase difference of the waves involved. This in turn depends upon the **path difference** between the waves involved. How far a wave has travelled determines its phase position, so if different waves individually travel different distances, they may be out of phase. We shall return to these ideas in chapter 3.2.

Classifying waves

Waves can be classified into progressive and **stationary** waves. Progressive waves are waves in which the positions of its peaks and troughs are moving. These waves are important because of their properties of 'action at a distance' – in other words, their properties as energy carriers. (Think of the way a pebble thrown into the water causes a water wave to spread out over the surface of a pond, carrying energy with it as it does so.)

Stationary waves are also known as **standing waves**. The words *stationary* and *standing* refer to the fact that the wave is not a progressive one – the positions of the peaks and troughs in it are not moving. Their importance lies in oscillations that occur in many systems, including stretched strings and in columns of air, and in structures such as bridges and vehicle components. They also provide a vital tool for understanding the behaviour of electrons and other subatomic particles.

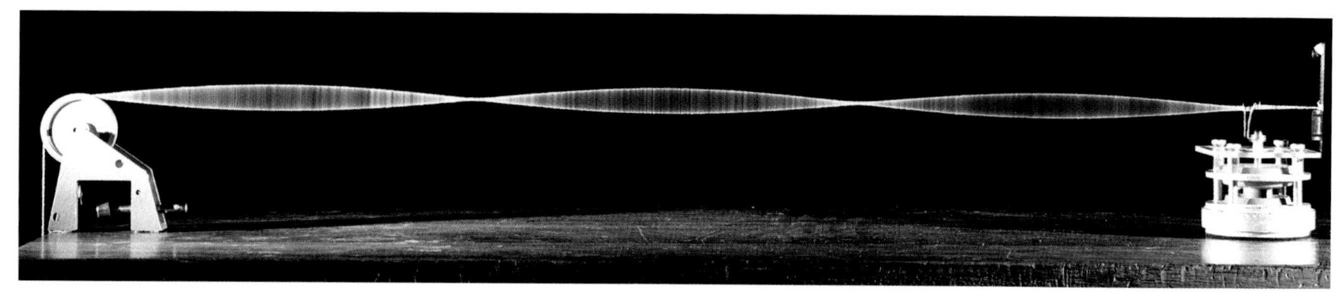

fig. 3.1.13 Standing wave on a vibrating string. A standing wave, also known as a stationary wave, is a wave that remains in a constant position. If the medium (here the string) is not moving with respect to the wave, then the standing wave is a result of interference between two waves travelling in opposite directions, in this case waves reflecting from either end of the string.

Stationary waves

An example of a stationary wave is the wave set up when the string of a guitar is plucked. When the string is plucked, it vibrates. Parts of the string vibrate to and fro, while other parts do not move. At its simplest, the string vibrates in a single loop, with a stationary point at either end and a point of maximum oscillation in the middle (**fig. 3.1.14**).

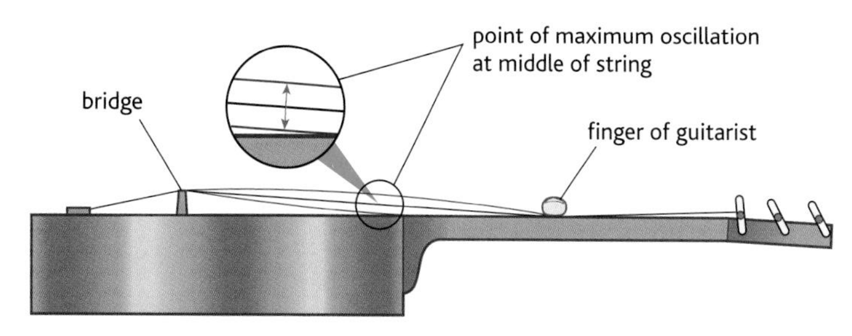

fig. 3.1.14 This diagram of a guitar string vibrating illustrates the simplest way in which a string may vibrate – called its **fundamental** mode. The points of no oscillation are called **nodes**, while the point of maximum oscillation is an **antinode**.

Making stationary waves on a stretched string

Transverse stationary waves may be investigated using a length of rubber cord in the apparatus shown in **fig. 3.1.15**.

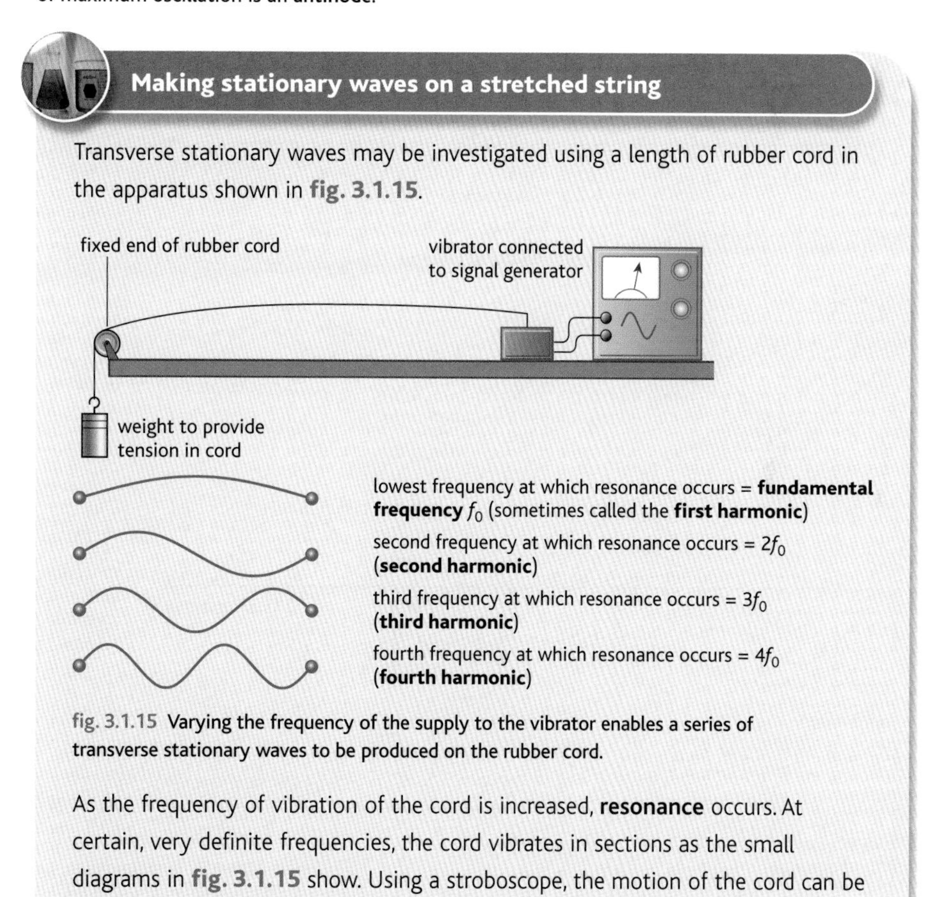

fig. 3.1.15 Varying the frequency of the supply to the vibrator enables a series of transverse stationary waves to be produced on the rubber cord.

As the frequency of vibration of the cord is increased, **resonance** occurs. At certain, very definite frequencies, the cord vibrates in sections as the small diagrams in **fig. 3.1.15** show. Using a stroboscope, the motion of the cord can be frozen or slowed down, so the up and down motion of the cord along its length can clearly be seen. This investigation can also be done by making waves along a slinky spring using your hand as the source of vibrations. The fact that stationary waves can only be set up at certain frequencies is quickly and easily established using this method.

The shape of the vibrating guitar string is due to the formation of a stationary wave on it. At the instant the string is plucked a progressive transverse wave travels along it from left to right. This wave is reflected back from the end of the string, and the result is a stationary wave. The principle of superposition can be used to explain how two progressive waves travelling in opposite directions can produce a stationary wave (**fig. 3.1.16**).

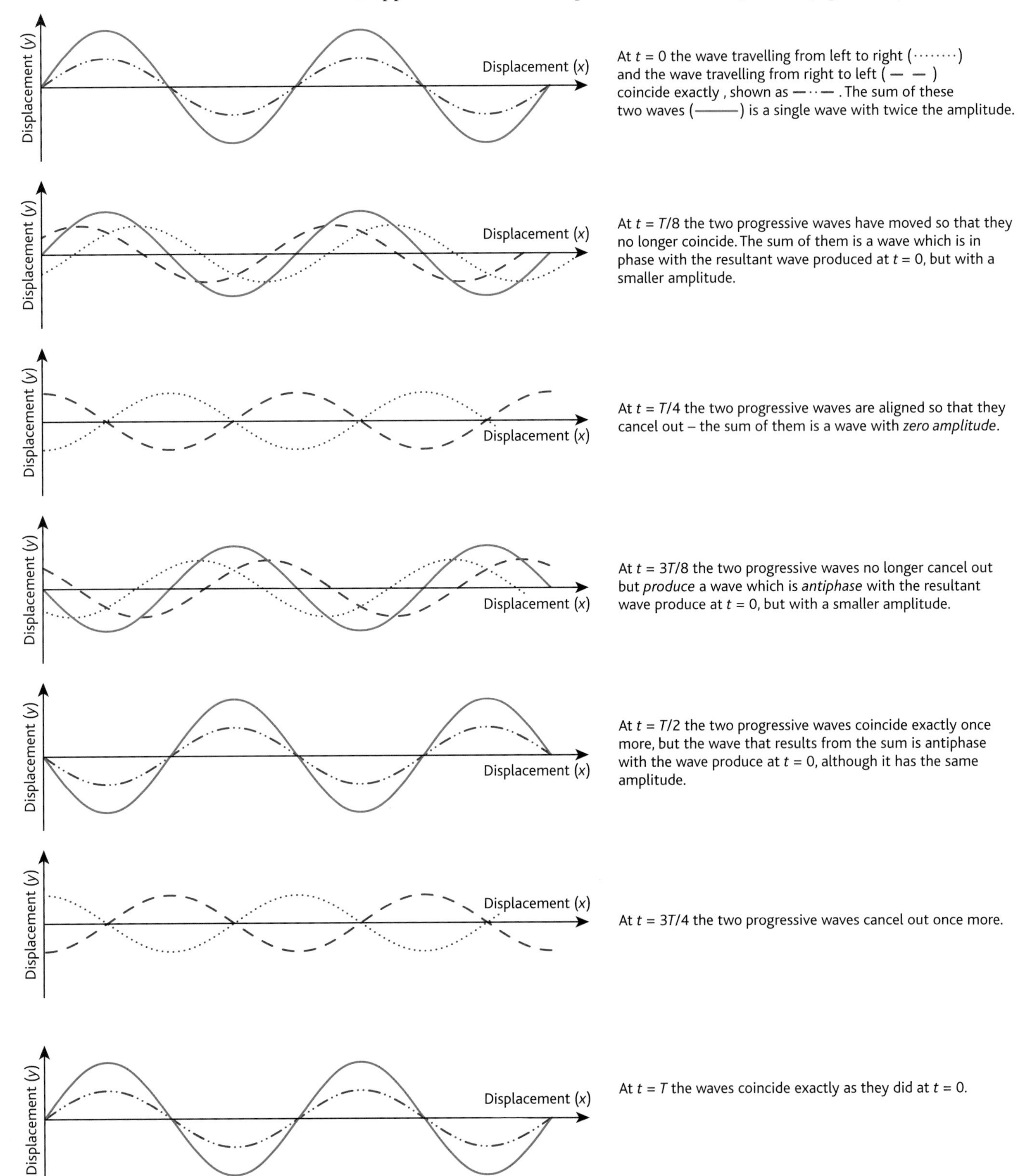

fig. 3.1.16 Two progressive waves moving in opposite directions produce a stationary wave.

For a string stretched between two fixed points (like the guitar string in **fig. 3.1.14**), the amplitude of the vibration of the string must be zero at either end, since it is fixed at these points. Given this restriction, it can be shown that the only waves that are possible on the string are those where:

$$\lambda = \frac{2l}{n}$$

where l = the length of the string and n is a whole number 1, 2, 3, etc.

This means that nodes will occur on the string at a distance of 0, $\lambda/2$, λ, $3\lambda/2$ from the end of it, and that neighbouring nodes are separated by a distance of $\lambda/2$, as are neighbouring antinodes. It also means that the fundamental mode of vibration of a string is an oscillation with a wavelength twice the length of the string, that is, $\lambda = 2l$. These points are shown in **fig. 3.1.17**.

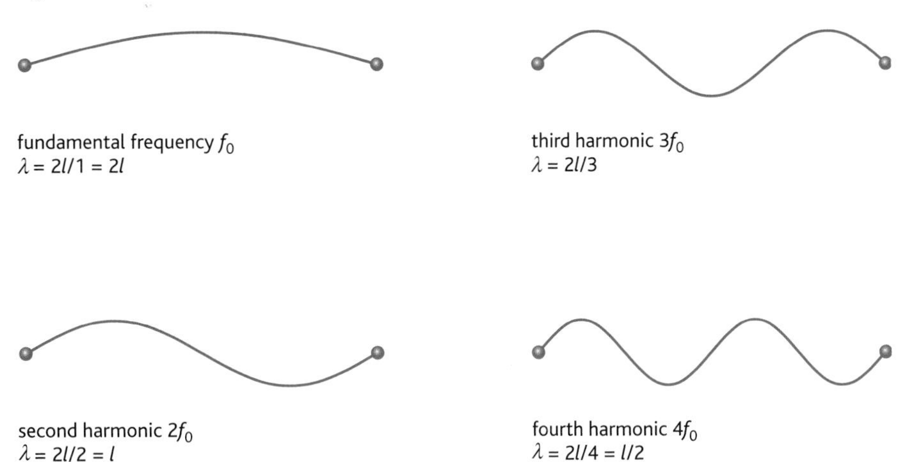

fig. 3.1.17 The first four modes of vibration of a stretched string. Notice that the distance between adjacent nodes or antinodes is always $\lambda/2$, and that there is always a node at each end of the string.

Questions

1 Sketch diagrams to show how a double bass string could vibrate at its fundamental frequency, and the mode of vibration producing its second harmonic. Label all nodes and antinodes, and annotate to show how the string length compares with the wavelength of the sounds.

2 **a** If the double bass in question 2 has an E string with a fundamental frequency of 41 Hz, calculate the frequency of the second and third harmonics.

b If the double bass E string is 106 cm long, what is the speed of sound in the string?

3 Certain factors, like weather, can cause a bridge to vibrate at its fundamental frequency, just like a guitar string. Under the right conditions, these vibrations can be very violent. Discuss why civil engineers would need to consider these factors when designing a bridge.

Reflection at the end of a string

The formation of a stationary wave on a string relies on the reflection of a progressive wave at the ends of the string. It also depends on the fact that such a reflection gives rise to a phase change of 180°.

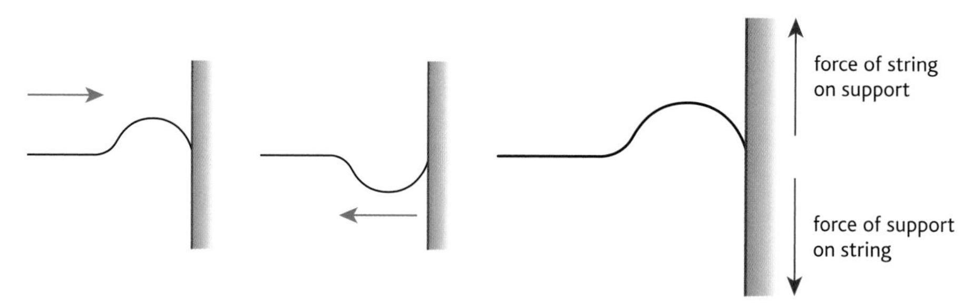

fig. 3.1.18 As the pulse reaches the end of the string the string exerts a force on the support. The support exerts a force equal in size but opposite in direction on the string, and so the pulse is reflected with a phase change of 180°.

This phase change on reflection occurs only where a 'hard' reflection occurs, where the string is connected to a massive solid object. Where the string is joined to a light object (another, lighter string, for example) reflection also occurs at the boundary between them, but with no phase change.

Other waves behave in the same way as waves on a string. For example, light waves are reflected with a phase change of 180° when travelling from a less dense medium to a more dense medium, but with no phase change at a more-dense-to-less-dense boundary.

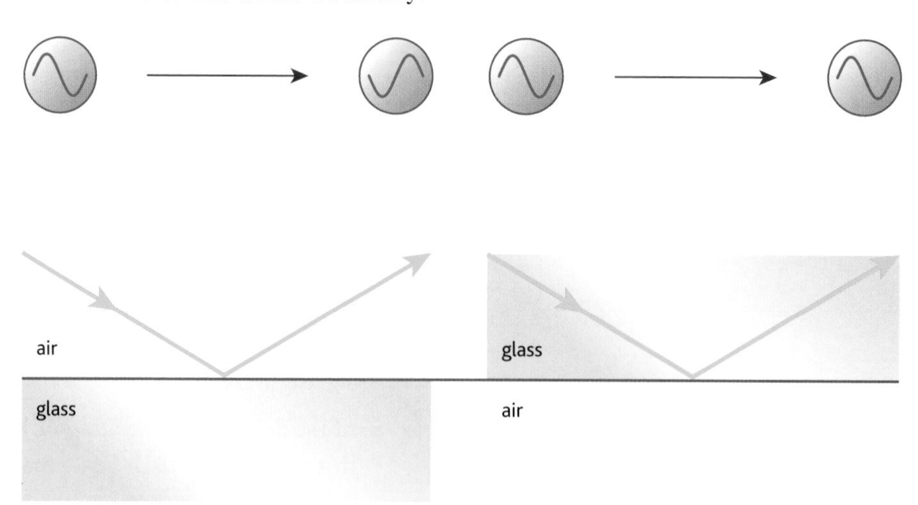

fig. 3.1.19 Phases changes in reflected light waves.

Harmonics

Many systems (not just stretched strings) have a set of natural frequencies of vibration. The set of vibrations associated with a particular system (whether it is a string or air column in a musical instrument, a tall skyscraper or the wing of a jet airliner) is called its harmonics, and the second and higher harmonics are sometimes (especially in the case of musical instruments) referred to as overtones.

fig. 3.1.20 The same note played on a piano and a saxophone will sound very different.

The note played on a piano is a result of the vibration of a string set in motion by being hit with a soft felt hammer. The way in which any string vibrates is governed by the way it is set in motion (basically where and how it is hit). This is also true for any other instrument – the saxophone, for example, has a vibrating air column instead of a string. The fundamental mode of vibration has the largest amplitude of any mode, and determines the pitch of the note produced. It is the combination of the fundamental mode of vibration coupled with the particular overtones for an instrument that determine the quality or timbre of a note, and that results in the difference in sound between two different instruments playing a steady note of the same pitch.

The way the loudness of the note produced by a musical instrument changes also helps to determine why we hear the notes produced by different musical instruments as being different (**fig. 3.1.21**).

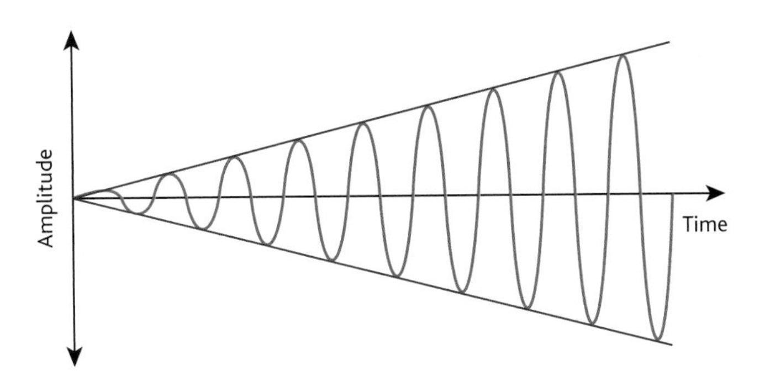

The rate at which the amplitude of the vibrations increases at the start of a note is sometimes called the **attack**.

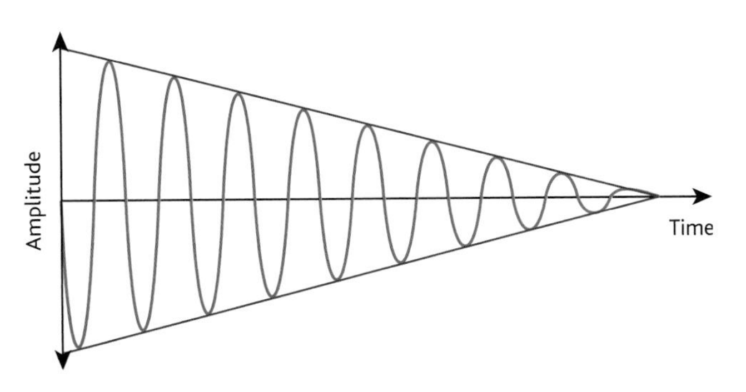

The term **decay** is sometimes used to describe the rate at which the amplitude of the vibrations dies away.

fig. 3.1.21 Attack and decay of a sound.

Finding the frequencies of the harmonics

To calculate the frequencies of the harmonics for a stretched string we can use the fact that the speed at which a transverse wave is propagated along a string is given by the relationship:

$$v = \sqrt{\frac{T}{\mu}} \qquad \text{(equation 1)}$$

where T is the tension in the string and μ is the mass per unit length of the string.

We also know that:

$$v = f\lambda \qquad \text{(equation 2)}$$

and that

$$\lambda = \frac{2l}{n} \qquad \text{(equation 3)}$$

Substituting into equation 2 the expression for λ from equation 3 gives:

$$v = \frac{2fl}{n}$$

or

$$f = \frac{n}{2l} \times v$$

Substituting the expression for v from equation 1 then gives:

$$f = \frac{n}{2l} \times \sqrt{\frac{T}{\mu}}$$

where n = 1, 2, 3, etc.

When a stretched string is set vibrating, all of these possible frequencies will occur at once, leading to a very complex mode of vibration. The complex vibrations produced by the fundamental and overtones of a singer's vocal cords and the air column in a trumpet can be seen clearly on the oscilloscope screens in **fig. 3.1.22**. These complex waveforms have been produced by the combination of a series of sinusoidally varying waves.

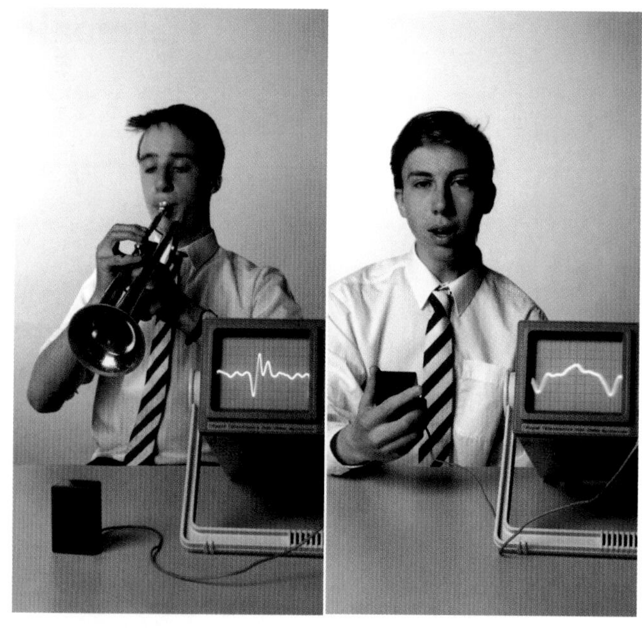

fig. 3.1.22 Oscilloscope traces of vibrations from a singer's vocal cords and a trumpet.

Worked examples

A piece of copper wire is fixed firmly at one end, and the other end is passed over a pulley and attached to a mass of 2 kg. The length of the wire between the fixed support and the pulley is 1.5 m. Separate measurements show that the mass per unit length of the wire is 20 g m^{-1}. What is the fundamental frequency of the wire if it is set in free oscillation?

We know that:

$$f = \frac{n}{2l} \times \sqrt{\frac{T}{\mu}}$$

so for n = 1, 2 and 3 we can substitute l = 1.5 m, T = 2.0 × 9.8 = 19.6 N and μ = 0.02 kg m^{-1}.

So:

$$f_0 = \frac{1}{2 \times 1.5} \times \sqrt{\frac{19.6}{0.02}} = \frac{1}{3} \times \sqrt{980} = 10.4\,\text{Hz}$$

Stationary waves – sound waves and microwaves

Both sound waves and microwaves (electromagnetic waves with a wavelength of around 3 cm – see chapter 3.3) can be used to investigate stationary waves, as **fig. 3.1.23** shows.

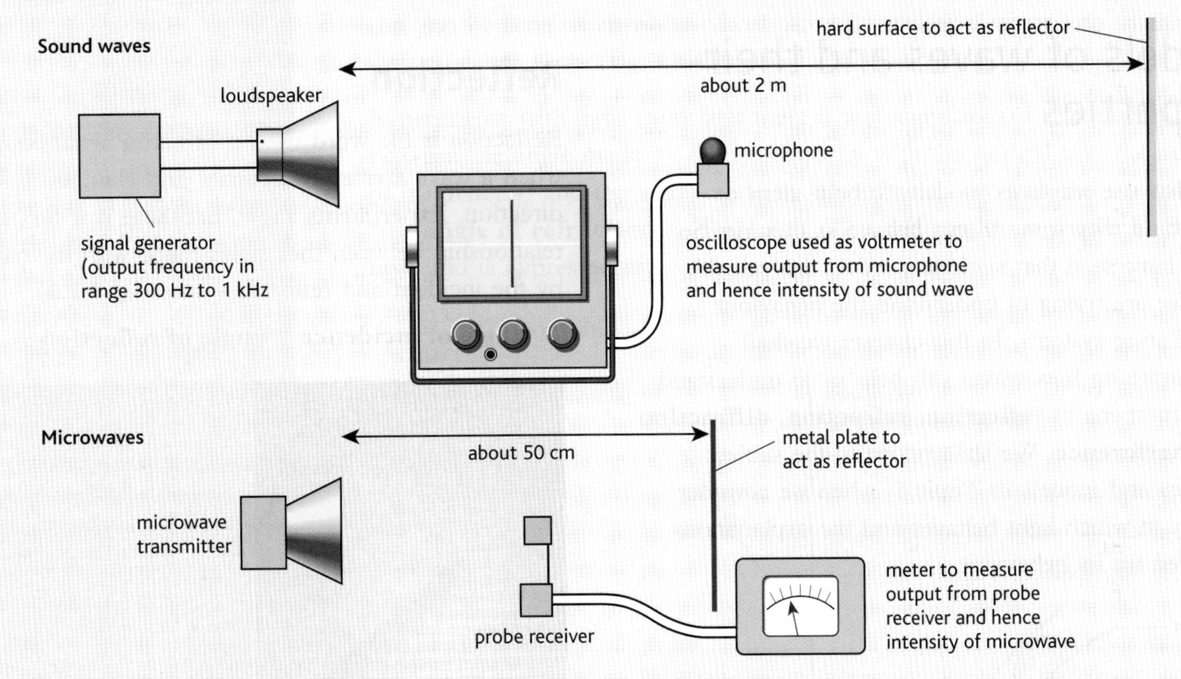

fig. 3.1.23 Investigating the production of stationary waves using sound waves and microwaves.

Moving the detector (the microphone in the case of sound waves, the probe receiver in the case of microwaves) along the line between the wave source and the reflector, alternating points of high and low signal strength can be found. These are the antinodes and nodes of the stationary wave. The distance between successive nodes or antinodes can be measured, and corresponds to half the wavelength.

If the frequency of the source is known or can be measured, this information can be used to calculate the speed of propagation of the two progressive waves that produce the stationary wave, from the relationship $v = f\lambda$. (Note: this is a suitable method for determining the velocity of sound waves in air but not the velocity of microwaves – why not?)

Questions

1 Explain how the vibrations of a loudspeaker cone show that sound waves are longitudinal and can be described in terms of the displacement of molecules.

2 In an experiment to set up a standing microwave, successive maximum readings (antinodes) were detected by the probe receiver every 1.4 cm.

  a What was the wavelength of the microwaves used in this experiment?

  b What was their frequency?

3 Use the idea that a hard reflection, such as that at the end of a guitar string, causes a phase change of 180° to explain why all stationary waves produced on guitar strings must always have a node at both ends.

4 Describe how you could set up an experiment to prove that different musical instruments playing the same pitched note are producing sounds with the same basic frequency, despite the instruments sounding different.

Worked example

A ray of light enters a pond at 30° to the horizontal. What is its direction as it travels through the water?

A ray travelling at an angle of 30° to the horizontal will make an angle of (90 – 30) = 60° to the normal. So:

$$\frac{\sin i}{\sin r} = \frac{\sin 60°}{\sin r} = 1.33$$

$$\text{or } \sin r = \frac{\sin 60°}{1.33} = \frac{0.866}{1.33} = 0.651$$

$$r = \sin^{-1}(0.651) = 41°$$

So the ray travels at an angle of 41° to the normal (or (90 – 41)° = 49° to the horizontal).

Wave speed and refraction

The change in direction, or refraction, that occurs when a wave enters a different medium is due to a change in the speed of the wave. The amount of refraction depends on the amount the speed changes. You have seen that the refractive index is a measure of how much a ray bends in moving from one medium to another. It is also equal to the ratio of the speeds in the two media.

$${}_1\mu_2 = \frac{\text{speed in medium 1}}{\text{speed in medium 2}} = \frac{v_1}{v_2}$$

The ratio of wave speeds is the refractive index for medium 2 with respect to medium 1 (i.e. it applies to rays entering medium 2 from medium 1), and is usually written as ${}_1\mu_2$. For rays entering medium 1 from medium 2 the ratio is reversed, and so:

$${}_2\mu_1 = \frac{v_2}{v_1} = \frac{1}{{}_1\mu_2}$$

Worked example

A ray of light enters a diamond. If it travels at $c = 3 \times 10^8\,\text{m s}^{-1}$ in air, how fast does the light travel in the diamond?

$${}_1\mu_2 = \frac{v_1}{v_2}$$

From **table 3.2.1**, we know that the refractive index for light entering diamond from air is 2.42.

$$2.42 = \frac{3 \times 10^8}{v_2}$$

$$v_2 = \frac{3 \times 10^8}{2.42} = 1.24 \times 10^8\,\text{m s}^{-1}$$

Measuring refractive index

Sometimes the refractive index of a material needs to be known very accurately. For example, the glass used to make spectacle lenses must have a precisely known refractive index if the lens grinder is to match the lens shape to the exact power needed for a person's eye prescription.

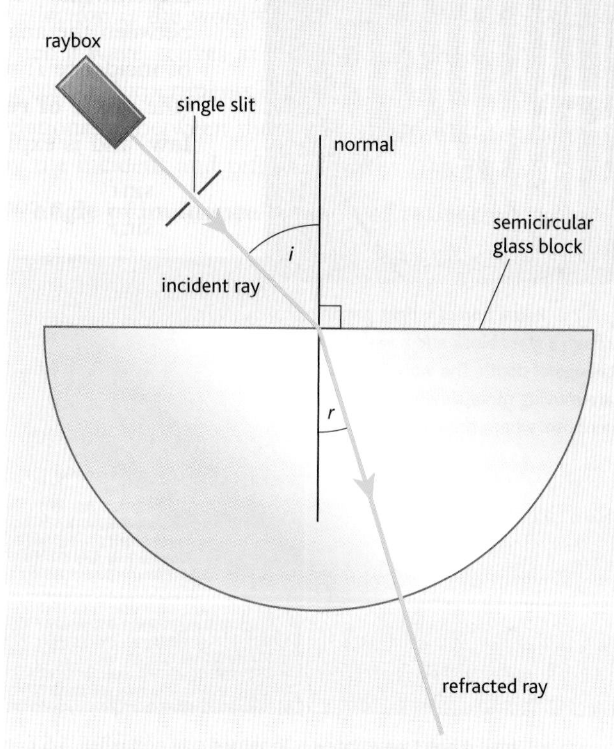

fig. 3.2.6 **Experimental setup for measuring refractive index.**

The angle of incidence i can be varied from 0° to 90° and the corresponding angle of refraction r measured. Because the light exits the glass at 90° to the boundary (it comes from the centre of the circle of which the curved edge is a part), there will be no further bending.

The results can be analysed by plotting $\sin r$ against $\sin i$. The straight line produced will have a gradient equal to $1/{}_1\mu_2$.

$${}_1\mu_2 = \frac{\sin i}{\sin r} \text{ so } \sin r = \frac{\sin i}{{}_1\mu_2}$$

Comparing this with the equation for all straight lines:

$$y = mx + c$$

shows that the y-intercept should be zero and that the refractive index will be given by:

$${}_1\mu_2 = \frac{1}{\text{gradient}}$$

Total internal reflection

When a light ray is refracted as it moves into a less optically dense medium and speeds up, it will be refracted away from the normal, according to Snell's law. You can investigate this by shining a ray of light through a semicircular glass block and taking measurements of the angles inside and outside the glass–air boundary as the ray tries to escape from the flat side of the block.

As the angle of incidence inside the glass block gradually increases, the angle of refraction becomes larger and larger, always being greater than the angle inside the block. A time comes when the escaping ray leaves at 90° and is effectively passing exactly along the flat face of the glass block. At this point, the internal angle to the normal is called the critical angle. If you increase the internal angle any further, the ray no longer leaves the glass. It reflects inside, following the law of reflection. In this case there is no refraction – the ray has been totally internally reflected. This is called total internal reflection.

Worked example

A ray of light reflects from a fish up towards the surface of a pond at an angle of 20° to the horizontal. What is the critical angle for a ray of light travelling from water into air? What will happen to this ray from the fish?

The refractive index from air into water, ${}_{air}\mu_{water}$, is 1.33, so from water into air the refractive index is:

$${}_1\mu_2 = \frac{1}{{}_2\mu_1} = \frac{1}{1.33} = 0.75$$

A ray travelling at an angle of 20° to the horizontal will make an angle of (90 – 20) = 70° to the normal.

The critical angle will be when the refracted ray leaves at 90° to the vertical.

$${}_1\mu_2 = \frac{\sin C}{\sin 90} = 0.75$$

$\sin 90 = 1$ so $\sin C = 0.75$

$C = \sin^{-1}(0.75) = 48.6°$

The critical angle is 48.6°.

The ray meets the water surface at an angle greater than the critical angle, so the ray will be totally internally reflected back into the pond water at the same angle of 20° to the horizontal.

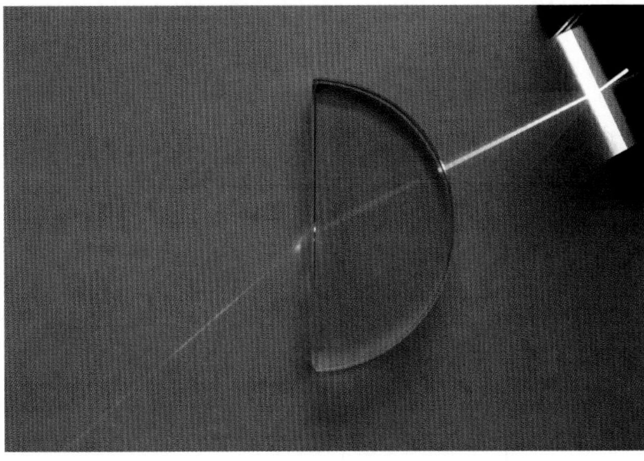

fig. 3.2.7 Investigating total internal reflection.

Partial reflection

It may not be obvious when you look at yourself in a mirror, but in most cases of reflection, not all the incident light is reflected. In general, at any boundary between media, some wave energy passes across the boundary – it is **transmitted** – whilst some of the energy is reflected. You may have noticed this when trying to look under a water surface which is reflecting a lot of light.

Questions

1. Use the data in **table 3.2.1** to find the critical angles for diamond, ice and liquid benzene.
2. How fast does light travel in crown glass? (See **table 3.2.1**.)
3. Describe an experiment which could be undertaken to measure the refractive index between glass and water. Include details of what measurements would be taken and how the results could be analysed to find a value for ${}_1\mu_2$.

Diffraction and interference

fig. 3.2.8 Light waves are not appreciably diffracted by a person, hence the crisp shadow formed on a sunny day. They *are* appreciably diffracted by the point of a pin, however.

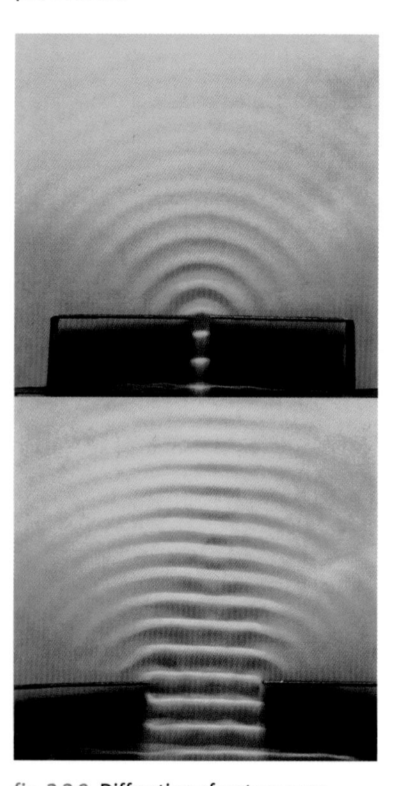

fig. 3.2.9 Diffraction of water waves occurs when the gap is similar in size to the wavelength, but not when it is considerably larger.

Diffraction

The term diffraction is used to describe the interaction between waves and solid obstacles. When a wave passes through a gap or round an object it may be deviated from its path. The amount of diffraction that occurs depends on the relationship between the size of the gap or the object and the wavelength of the wave – when the two are similar in size, substantial diffraction occurs (**fig. 3.2.8** and **fig. 3.2.9**).

Investigating the diffraction of light

The apparatus shown in **fig. 3.2.10** may be used to investigate the diffraction of light. The light from the laser is diffracted by the slit, which should be placed about 1 m from the screen (a piece of thick paper or white card). The width of the slit is slowly reduced, starting at about 2 mm, until a diffraction pattern is seen on the screen. Marks can be made on the screen to record the spacing of the bands in the diffraction pattern.

Note that you should *never* view laser light directly – the diffraction pattern should be viewed by reflection of the light from an opaque screen

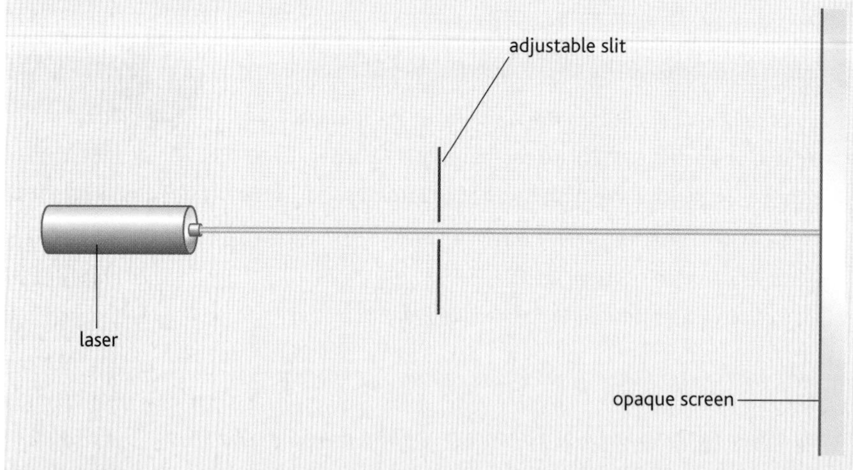

fig. 3.2.10

This investigation relies on the properties of laser light, and requires modification if an ordinary light source is used. If a laser is not available a bright lamp and single slit may be used to illuminate double slits, and a translucent screen must replace the opaque screen. It will also be necessary to work in a darkened room.

HSW The nature of electrons

Following the work on the particle nature of electromagnetic waves, Louis de Broglie suggested that it was possible that particles like electrons might have wave properties. The wave-like nature of electrons is fundamental to our present understanding of the behaviour of electrons in atoms, as well as the behaviour of all the other 'particles' found in nature – protons, neutrons, neutrinos and so on. In 1924, de Broglie stated that the wavelength associated with a particle (called the de Broglie wavelength) is inversely proportional to its momentum. For an electron, this means that the faster it moves, the smaller its associated wavelength.

The wave properties of the electron were subsequently confirmed independently by the American Clinton Davisson and by George Thomson in England in 1927, showing the diffraction of electrons by crystals (**fig. 3.2.11**). For significant diffraction, the wavelength of the wave must be of a similar size to the gap the wave is passing through. Thus, to achieve electron diffraction by passing them through the gaps between atoms in a crystal, the electron wavelength must be of the order of 10^{-10} metres. This means the electrons must be travelling at 2–3% of the speed of light.

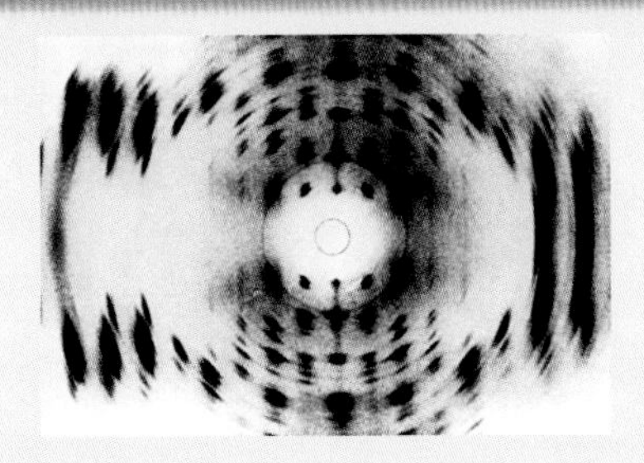

fig. 3.2.11 **Under the right conditions, electrons can exhibit wave properties like diffraction.**

Three years after de Broglie had come up with his wavelength relationship for particles, the Davisson–Germer experiment aimed a beam of electrons at a crystal of nickel. The diffraction pattern produced matched exactly with similar experiments using X-rays and confirmed the wave nature of electrons. In 1929, de Broglie was awarded the Nobel Prize for Physics 'for his discovery of the wave nature of electrons'. In 1937, Davisson and Thomson shared the Nobel Prize for Physics 'for their experimental discovery of the diffraction of electrons by crystals'. George Thomson was the son of Joseph John Thomson, who had discovered the electron as a charged particle and gained the Nobel Prize for Physics in 1906.

fig. 3.2.12 Two circular waves (produced by dippers) on a ripple tank overlap to produce a distinctive interference pattern.

Interference

Interference occurs when waves overlap each other to produce a pattern where the waves reinforce each other in some places and cancel each other out in others (**fig. 3.2.12**). The principle of superposition which we met in chapter 3.1, and that we saw explain the behaviour of waves crossing each other, also provides an explanation for interference.

Interference effects using light were first demonstrated at the beginning of the nineteenth century by Thomas Young. Young's experiment used two narrow slits to produce an interference pattern from light. Although Young's results provided strong evidence for the wave nature of light, physicists did not accept that light had wave-like properties until much later on, as we shall see in Topic 5. **Fig. 3.2.13** shows how Young's experiment can be repeated.

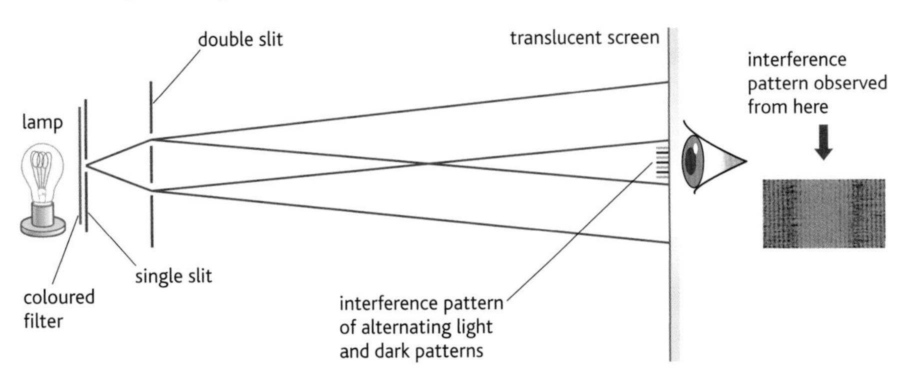

fig. 3.2.13 Young's slits experiment.

Similar results can be obtained using sound, which is a **longitudinal** wave (**fig. 3.2.14**). In this case two loudspeakers driven by a single signal generator can be used, ensuring that the two sources are in phase (i.e. the cones of both loudspeakers move forwards and backwards together). For a sound with a pitch of 1000 Hz, the two loudspeakers should be placed about 2 m apart. An observer walking along line AB will hear the intensity of the sound rise and fall, as constructive and destructive interference alternate.

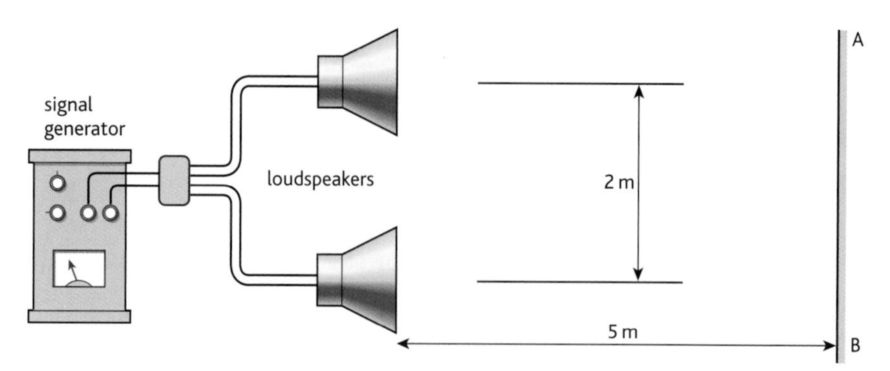

fig. 3.2.14 Demonstrating interference in sound waves.

Phase difference and path difference

The production of an interference pattern of light and dark bands by two parallel slits can be explained by thinking about the **phase** of the waves arriving at the screen (**fig. 3.2.15**). If two waves arrive in phase they reinforce each other and a bright area is seen. This is **constructive interference**. If two waves arrive in antiphase (180° out of phase) they cancel each other out and a dark area is seen. This is **destructive interference**.

Fig. 3.2.16 shows how the pattern of light and dark bands (often referred to as **fringes**) arises. In this diagram the distance between the two slits A and B is exaggerated and the distance D from the slits to the screen reduced in order to make the situation clearer.

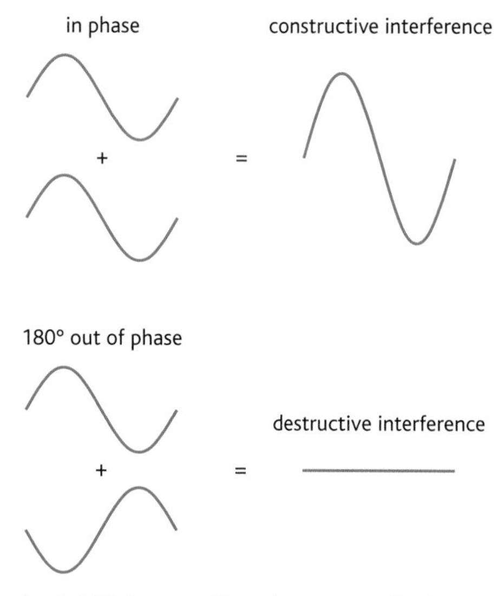

fig. 3.2.15 Superposition of waves resulting in constructive and destructive interference.

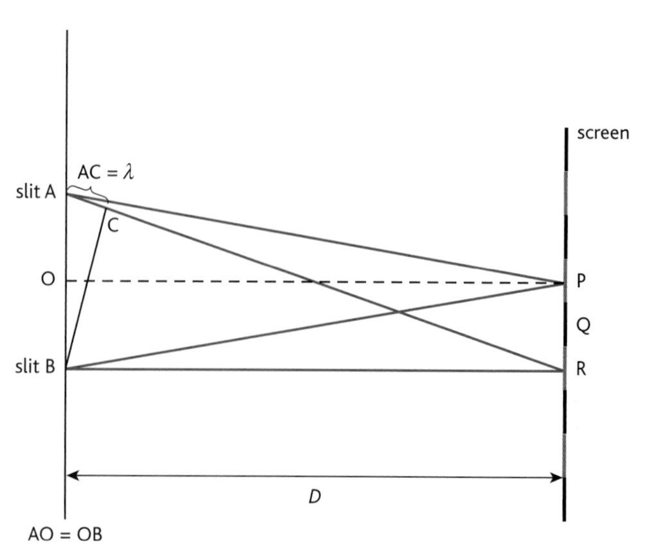

fig. 3.2.16 Path differences cause interference.

Since the light passing through the slits is from the same source, the light waves leave slits A and B in phase. Since AP = BP, the waves must arrive at P in phase, so constructive interference occurs here and a bright area is seen.

The distance AR is exactly one wavelength more than the distance BR, so the waves also arrive at R in phase, resulting in a bright area here also.

The distance AQ is exactly half a wavelength more than the distance BQ, so the waves arrive at Q in antiphase, resulting in a dark area.

The production of an interference pattern by two wave sources can only be observed if the sources are **coherent**. For waves to be coherent there must be a constant phase difference between the two waves. For example, the waves may always be in phase with each other, in which case there is a zero phase difference. Alternatively, one might always be a quarter of a cycle behind the other one. The phase difference itself doesn't matter. As long as it remains constant all the time, the waves are coherent and can show us an interference pattern. If the phase difference keeps changing, then the waves are **incoherent**. Incoherent wave interference will sometimes be constructive, and sometimes destructive. As the frequencies are generally very high – yellow light has a frequency of 550 thousand billion hertz – the changes in interference at any point we try and observe will be so rapid that we will not be able to see any clear outcome.

HSW Radio interferometers

The atmosphere does not absorb much of the radio wave energy that hits the Earth from stars that emit radio frequencies. This means that radio telescopes based on the ground can be used very successfully. The larger a radio telescope can be made, the better its resolution, which is a measure of the fine detail it can observe. A system that uses the ideas of interference has been used by radio astronomers to artificially increase the effective size of their radio telescopes.

fig. 3.2.17 **The Very Long Baseline Array consists of ten radio dish telescopes spread over 5000 miles across the USA.**

With two or more radio telescopes linked together, computer processing of the signals they receive can highlight very slight differences in phase at each place. These phase differences are caused by the very slight path differences from the source to each of the radio dish locations. By analysing these interference patterns, the group of dishes can act like one enormous radio telescope, improving the resolution significantly.

Questions

1 Copy the diagram, which shows wavefronts approaching a gap. From the wavefront that is actually within the gap, use Huygens' construction to show the secondary wavelets that would be produced from several points on this wavefront. Continue the diagram to show how Huygens' construction could account for diffraction through a small gap.

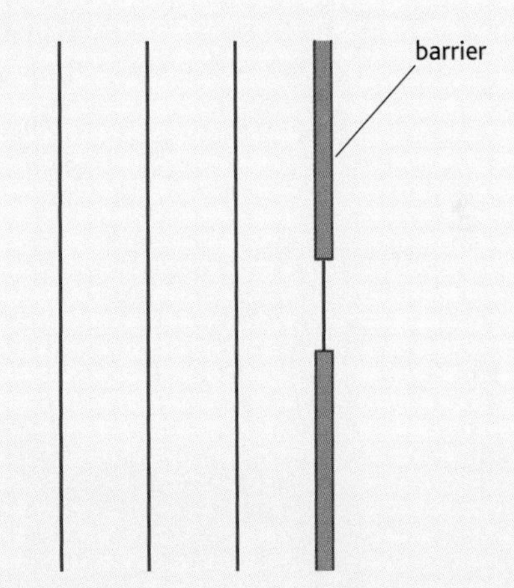

fig. 3.2.18

2 In driving due north along a straight road a driver notices that the radio station she is listening to gets louder and quieter as she drives along. Explain this, if she can see two separate radio transmitters in the distance, to the west of her car.

3 Why was the independent confirmation by separate experiments such an important part of the development of the idea of the existence of electron waves?

Polarisation

In contrast to superposition, which is something that *all* waves exhibit, the phenomenon of **polarisation** is something that only transverse waves show. **Fig. 3.2.19** shows the some of the planes in which the oscillations in a transverse wave may occur. A wave in which the oscillations take place in a number of planes is called **unpolarised**, while a wave in which the oscillations occur in one plane only is said to be **plane polarised** in that direction.

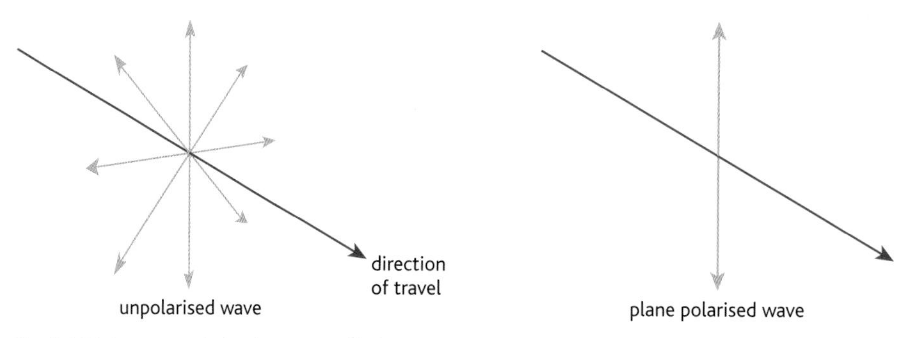

fig. 3.2.19 In an unpolarised wave, oscillations may occur in any plane, while in a plane polarised wave they occur in only one plane.

Electromagnetic waves are transverse waves, and so may be polarised. This property can be very useful. **Fig. 3.2.20** shows how polarisation can be used in transmission of TV signals. The main transmitter has its aerial in the horizontal plane, so signals from here are horizontally polarised. The relay station boosts the signal for areas where reception would otherwise be poor. The aerial of this transmitter is vertical, so signals from here are vertically polarised. This means that these signals cannot interfere with the signals from the main transmitter.

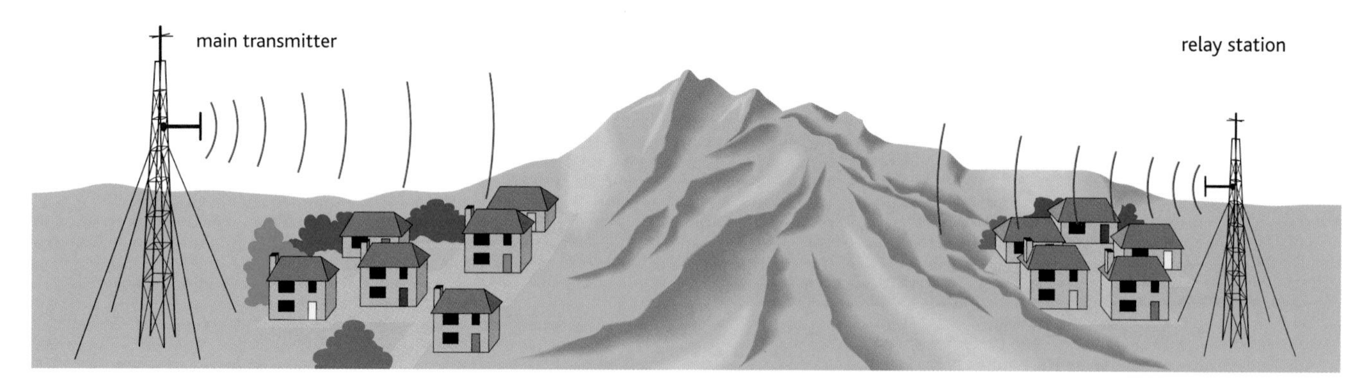

fig. 3.2.20 Polarisation may be used so that signals from different television transmitters at similar frequencies do not interfere with each other.

Light from the Sun or from an electric filament lamp is unpolarised because the waves are emitted at random from the atoms of the object. Such light may be plane polarised by passing it through a polarising filter. In polaroid, long molecules of quinine iodosulfate are lined up so that only light waves oscillating in one particular plane can pass through. This property is used in polaroid sunglasses (**fig. 3.2.21**). Two pieces of polaroid that have the molecules aligned at 90° to each other will not allow any light through.

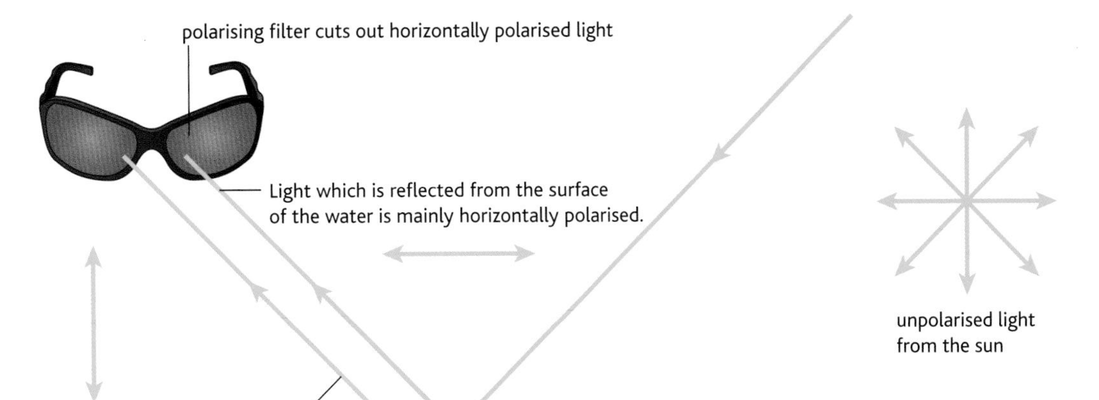

fig. 3.2.21 Polaroid sunglasses reduce glare from the water surface by cutting out horizontally polarised light.

fig. 3.2.22 The molecules of life itself react to plane polarised light – all but the simplest biological molecules exhibit optical activity.

Investigating the polarisation of light

light sensor connected to laptop
reference mark
card mount marked in degrees
polaroid
sugar solution
polaroid
light source

fig. 3.2.23

The transmission of light through polarising filters can be investigated in the school lab. Firstly, the amount of light transmitted through a pair of polaroid sheets held at various angles to one another will demonstrate the effects of these filters. You can then extend this to observe the rotation of the plane of polarisation caused by different concentrations of sugar solutions. This is called polarimetry and is used in the confectionery industry to measure accurately the concentration of solutions being used to manufacture sweets.

Questions

1 Explain why the orientation of a television aerial could make a big difference to picture quality.

2 Why can't sound waves be used to test the concentration of sugar solutions by polarimetry?

3.3 Light and sound

Light as a wave

So far in this topic we have looked at both stationary and progressive waves and the properties they possess. In this chapter we shall concentrate on the two wave motions that are most directly relevant to the lives of those of us who are blessed with the senses of sight and hearing – light and sound.

For the rest of this topic we shall treat light as a wave motion, since it is its wave-like properties that are of interest to us. However, remember that it does behave differently under different circumstances – we shall investigate these in Topic 5.

Electromagnetic waves

In the 1860s, the Scottish mathematician and physicist James Clerk Maxwell produced a theory to explain light in terms of electric and magnetic fields, using work on magnetism and electricity previously begun by Michael Faraday. Maxwell reasoned that as a moving magnet could induce a current to flow in a wire and that a current flowing in a wire was also the cause of a magnetic field, it ought to be possible to produce some sort of combination of an electric field and a magnetic field which was 'self-sustaining'. Using some fairly complex mathematics, Maxwell showed that light can be pictured as a combination of a transverse 'electric wave' and a transverse 'magnetic wave' moving through space – an **electromagnetic wave**.

Maxwell suggested that the electromagnetic (EM) wave is caused by an oscillating electric charge which sets up an oscillating electric field. This oscillating charge is often an electron. Light waves are the result of electrons moving within atoms, while radio waves are the result of electrons moving in wires (transmitting aerials).

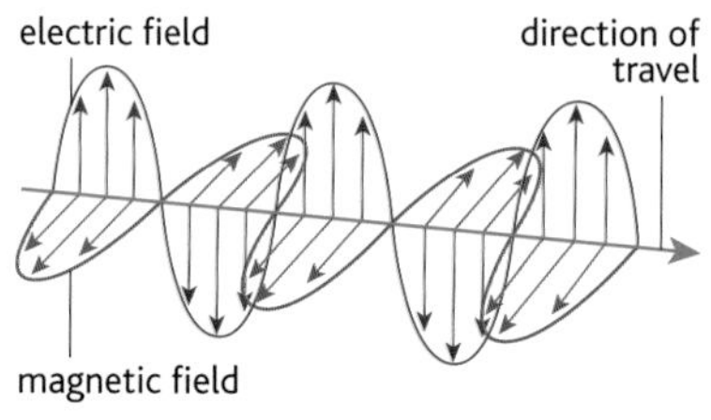

fig. 3.3.1 **James Clerk Maxwell's model of an electromagnetic wave.**

The oscillating electric field causes an oscillating magnetic field. The two fields are in phase, the electric field oscillating in one plane (vertical in **fig. 3.3.1**) while the magnetic field oscillates at right angles to it. The plane of polarisation of the wave is the plane in which the electric field oscillates.

By modelling the way that the electric field and the magnetic field varied with one another according to equations describing electric and magnetic fields, Maxwell was able to show that light waves:

- travel through a vacuum with a velocity of $3 \times 10^8\,\mathrm{m\,s^{-1}}$, agreeing with the measurement of the speed of light made by Fizeau in 1849
- are part of a larger family of waves with a large range of wavelengths.

The large family of waves predicted by Maxwell forms the **electromagnetic spectrum**. All the waves in this family travel through a vacuum at the same speed, although the family covers the vast range from radio waves with wavelengths of several kilometres to gamma rays with wavelengths of as little as 10^{-16} m. As with all waves, electromagnetic waves can be described by the wave equation:

speed = frequency × wavelength

The speed of electromagnetic waves takes the symbol c, and in a vacuum is always $3 \times 10^8 \text{ m s}^{-1}$ for all parts of the electromagnetic spectrum.

Thus, in a vacuum:

$$c = f\lambda. = 3 \times 10^8 \text{ m s}^{-1}$$

Fig. 3.3.3 gives some more details of the family of electromagnetic waves.

HSW Maxwell's work on electromagnetic waves

Maxwell used earlier work by two distinguished physicists in his model of light waves – Michael Faraday and André-Marie Ampère. Faraday and Ampère had set out laws relating the movement of electric charge (in the form of electric currents) to changes in magnetic fields. These laws, when applied to oscillating electric and magnetic fields showed that:

- an oscillating electric field and an oscillating magnetic field travelling through space can be self-sustaining (i.e. the oscillating electric field produces an oscillating magnetic field and vice versa)
- the two oscillating fields can be self-sustaining only if they are at right angles to each other and to the direction of propagation and are in phase
- the waves must travel through a vacuum with a fixed speed.

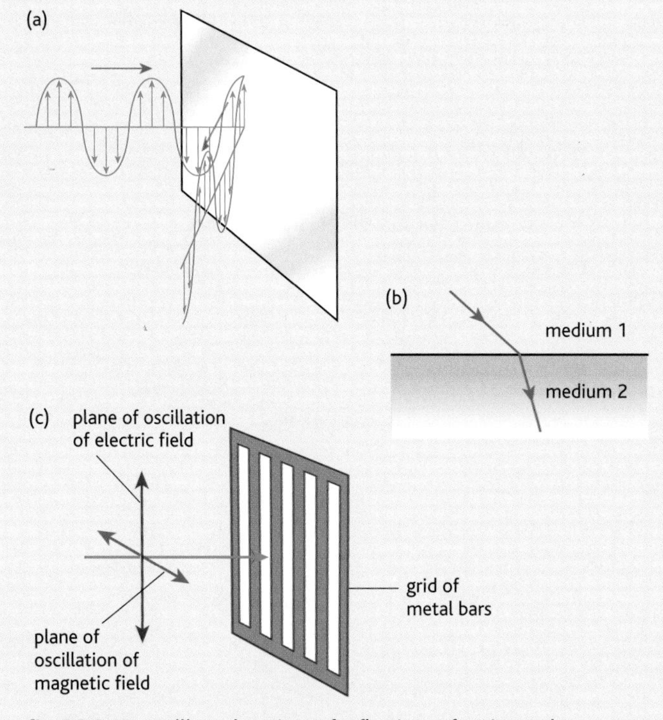

fig. 3.3.2 Maxwell's explanations of reflection, refraction and polarization in electromagnetic waves.

With such a model, explaining the behaviour of electromagnetic waves such as light and radio waves becomes a task of explaining the way in which the oscillating electric and magnetic fields interact with matter.

Maxwell's explanation for the reflection of light is that oscillating fields in the incident wave set up oscillations of electrons in the reflecting surface (**fig. 3.3.2a**). This leads to the absorption of the energy of the incident wave by the reflecting surface, which re-radiates it.

As we saw in chapter 3.2, refraction is explained by a change in speed as a wave enters a new medium (**fig. 3.3.2b**). Experiments with light show that the wave must slow down on entering a denser medium. Maxwell's mathematics about the movement of the electric and magnetic fields of the wave through different media led to the required decrease in speed.

Maxwell's explanation for polarisation is illustrated in **fig. 3.3.2c**. With the grid in this orientation, the oscillating fields in the wave can make the electrons in the metal bars oscillate in a vertical plane. This means that the energy of the wave is absorbed by the grid and re-radiated in all directions – that is, the wave cannot effectively pass through the grid. With the bars of the grid rotated through 90°, there is little absorption of the wave energy by the electrons in the bars and it passes through the grid unaffected. At intermediate angles some of the wave's energy will be absorbed and re-radiated with a plane of polarisation in the same plane as the orientation of the bars in the grid. This means that the plane of polarisation of an electromagnetic wave describes the plane of oscillation of the electric field in the wave.

The electromagnetic spectrum

Light is a small part of the electromagnetic spectrum, consisting only of the part to which our eyes are sensitive. This is deceptively simple however, for just as within the visible spectrum colours merge into each other so that it is difficult to tell where yellow finishes and orange begins, the visible spectrum merges at one end into the **infrared** and at the other into the **ultraviolet**, with no definite cut-off point. It is just the same in the rest of the electromagnetic spectrum, where the properties of the waves change gradually as their wavelength and frequency change.

Questions

1 Calculate the frequencies of the following:

 a an infrared wave with a wavelength of 4×10^{-5} m

 b red light at the long wavelength limit of human vision (700 nm)

 c a gamma-ray photon for which $\lambda = 6.5 \times 10^{-13}$ m.

2 Explain what previous knowledge James Clerk Maxwell had that he combined to come up with the idea of 'self-sustaining electromagnetic waves'

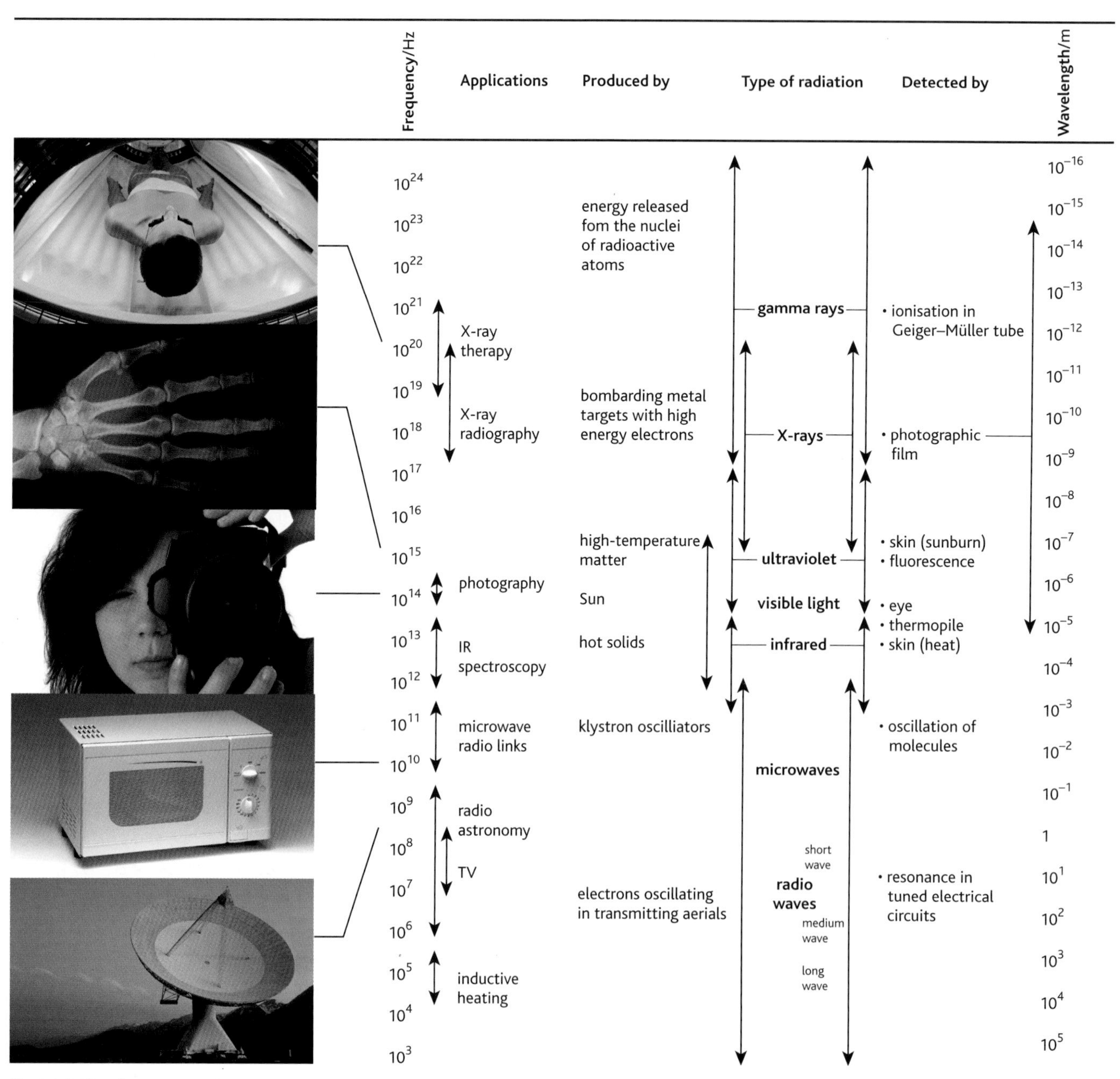

fig. 3.3.3 The electromagnetic spectrum.

Applications of electromagnetic waves

Some of the parts of the electromagnetic spectrum have very familiar uses, such as the microwaves that heat food in a microwave oven, or radio waves used in broadcasting. The nature of the uses to which humans have put electromagnetic waves depends on the way in which their transmission and reflection vary across the spectrum. For example, visible light will pass through glass, but infrared will not, so glasshouses become warm enough to help plants to grow. Light energy goes in, but heat radiation cannot escape.

Radio

Broadcast communications are the most obvious use for **radio waves**. These are particularly easy to generate as any alternating current can produce them. By careful arrangement of the aerial and the ac, we can produce radio waves of a chosen frequency. Since these will not interfere with those of another frequency, many radio stations can broadcast in the same area.

An upper layer in the atmosphere, the ionosphere, is electrically charged and will reflect and refract radio waves. This can be useful for long-distance broadcasting of radio signals. The waves travel in straight lines and would be lost by the curvature of the Earth, but with ionospheric reflection, radio broadcasts can get everywhere from one transmitter. However, this means that radio communications with satellites is not possible for those frequencies affected, which are generally in the range 3–300 MHz.

Microwaves

As the frequency of radio waves increases into the microwave region, they are not reflected by the ionosphere, and thus **microwaves** are used for communications with satellites. Microwaves have a shorter wavelength too, so they need a smaller aerial for transmitting or receiving. This makes them ideal for handheld communication devices such as mobile phones.

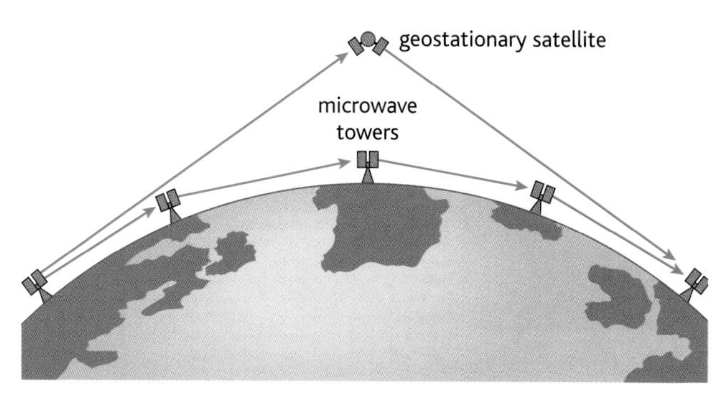

fig. 3.3.4 As microwaves travel in straight lines and are not affected by the ionosphere, they can only communicate in straight lines. Over long distances, this needs either a chain of repeater towers or satellite relays.

Infrared

Between microwaves and red visible light is the region called infrared (IR). These waves travel through glass in a very similar way to visible light and can use **total internal reflection** to be carried along optical fibres. They are used for wired communications (along optical fibres) along with visible light, to increase the number of available frequencies that can be used. This maximises the amount of data that can be transmitted simultaneously.

These electromagnetic waves are also the ones responsible for heat radiation. When you feel the warmth from a bonfire at a distance from it, that heat is being carried to you primarily by infrared waves. Similarly, toasters and grills use infrared to cook food.

Visible light

Human eyes can detect electromagnetic wavelengths from about 400 nm (violet) to 700 nm (red). This is an incredibly small section of the electromagnetic spectrum. However, our eyes are not as limited as you might think. Apart from a large range of radio wavelengths (which would need a metal aerial to detect) virtually the only wavelengths that naturally occur at the surface of the Earth are those we can see. Some birds and insects can also see in a range of ultraviolet wavelengths that we cannot see. Indeed, some birds have four types of colour receptors (cone cells) in their eyes, compared with three in humans. This not only gives birds a wider wavelength range but, when their brains combine the signals from four types of detector, their vision is very much richer than ours.

fig. 3.3.5 Visible light is totally internally reflected within glass. This allows images to be transported directly from one place to another. For example, a surveillance camera can send its images to a monitor in a nearby room.

Any application that includes our vision relies upon using visible light.

Ultraviolet

Ultraviolet (UV) waves have higher frequencies than visible light, ranging from about 7.5×10^{14} Hz up to 3×10^{16} Hz. Some of the higher-energy groups of ultraviolet waves are used in the treatment of sewage to kill bacteria. In the same way that ultraviolet can cause sunburn in humans, the energy of the ultraviolet waves ranging from 100 nm up to about 315 nm (UV-B and UV-C) can be used to destroy harmful organisms in wastewater. The DNA of various types of microorganisms, including bacteria and viruses, can be damaged enough to kill these organisms and make the wastewater safe for human consumption (**fig. 3.3.6**). This is a good way of recycling precious water as it does not add chemicals to the water. On the other hand it is very energy intensive, and energy is a similarly precious resource.

fig. 3.3.6 Ultraviolet light can be used to kill harmful organisms in wastewater.

X-rays

Very high energy electromagnetic waves can be produced by colliding high-speed electrons with a metal target. When these electrons decelerate rapidly, they give off energy as **X-rays**. This is how a hospital X-ray machine produces its electromagnetic waves. These waves will affect photographic film, which causes the cloudy picture you may be familiar with in medical X-rays. Recently, X-ray detectors have been developed that do not rely upon a chemical reaction. These can produce digital images of the X-rays detected without needing to use up a constant supply of photographic film. X-ray scanning can also be used in any industry in which the integrity of metal parts needs to be tested – aeroplane engines, for example. The part is not damaged by the test, but can be checked for cracks and other flaws. Dense materials, like metals, will absorb X-rays, and can produce a shadow image, just like a medical X-ray image. If the metal is cracked, more X-rays will pass through without being absorbed and this will show up on the image.

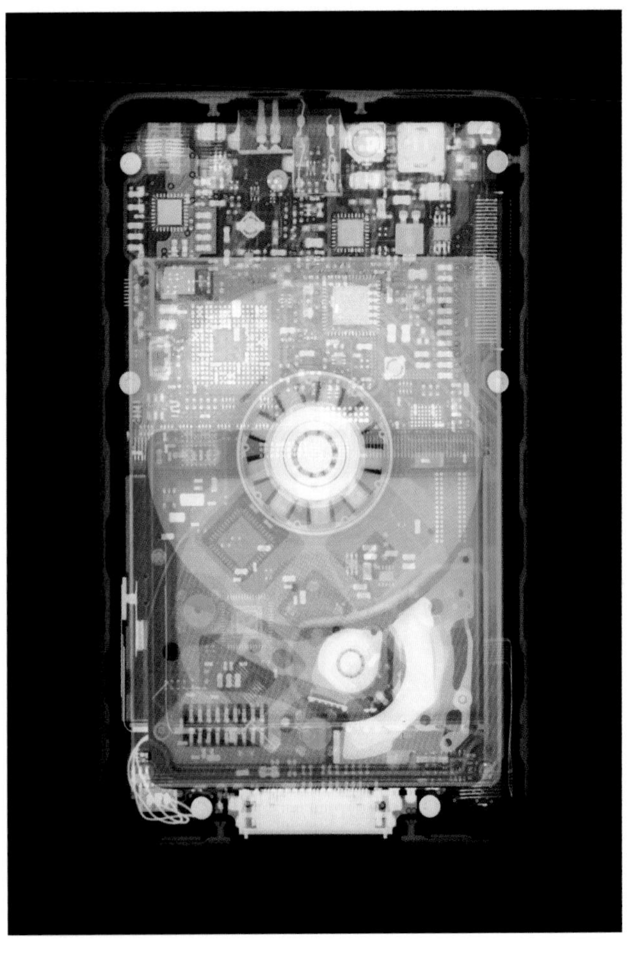

fig. 3.3.7 X-rays can show us inside metal objects without breaking them open.

HSW Astronomy and the electromagnetic spectrum

Stars produce electromagnetic waves in all parts of the electromagnetic spectrum. In some cases, different types of stars produce a particular type of electromagnetic wave that can be observed for a particular reason. For example pulsars, which are some of the most distant stars that can be observed, are very strong radio emitters. In fact, the emissions they produce are so strong that scientists have yet to work out how they can produce them. Pulsar emissions can be observed using radio telescopes on Earth. Visible light from the heavens can also be observed at ground level using an ordinary telescope. Most parts of the electromagnetic spectrum are absorbed by the atmosphere before reaching the ground and can thus only be observed by astronomers using satellite-based detectors. The development of satellites has opened up vast new areas of research in astronomy, as it is now possible to scan the skies at wavelengths that were previously invisible.

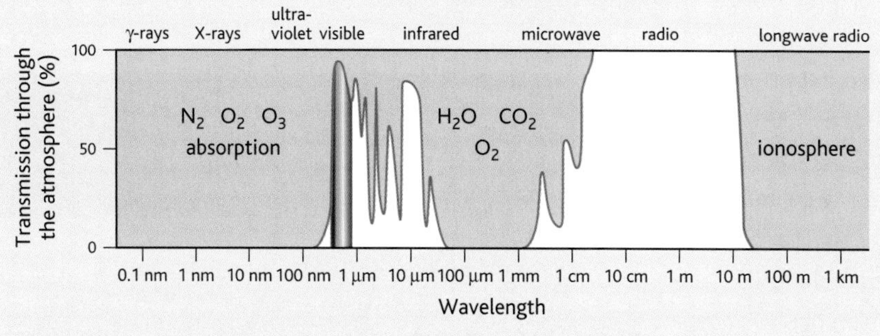

fig. 3.3.8 **The atmosphere absorbs most wavelengths of electromagnetic radiation, protecting life at the surface of the Earth from high-energy waves, but limiting the scope for astronomy.**

Some people claim that, apart from interesting knowledge, astronomers contribute nothing productive to society. These people feel that, for example, detecting radio wave pulses from spinning neutron stars is not a good way to use our resources. The brains of such clever scientists, along with their time, and the money and energy that they use in their research, might be better spent on medical research or agricultural improvements – things that could obviously and directly help people now. At present, global society is generally supportive of astronomers and we are willing to feed and clothe them in return for nothing more than knowledge. Should it be this way?

fig. 3.3.9 **Throughout history, societies have chosen to support the work of astronomers who seemingly contribute nothing in return. Why?**

Gamma rays

Generated by energy shifts inside the nuclei of atoms, **gamma rays** form the very highest energy EM waves. At lower energies, there is a crossover between the frequency ranges of X-rays and gamma rays, but the distinction comes from how the electromagnetic waves were produced. Most gamma-ray applications come from their property of being fatal to biological cells. They are used in the sterilisation of surgical instruments and soft fruits. Gamma rays are also used to kill cancerous cells in the body. This can be a dangerous procedure, as the interaction between healthy cells and gamma rays is not very good for the body.

Questions

1 a Calculate the wavelength range of radio waves that suffer reflection by the ionosphere.

b Explain why communications with satellites must use radio/microwaves with a wavelength shorter than 1 metre.

2 Explain why the human eye did not develop so that it could detect UV light with a wavelength of 100 nm.

3 Give a similarity and a difference between X-rays and gamma rays.

4 Explain, using diagrams if necessary, how X-rays could be used in a machine to produce images that could be scrutinised to detect flaws in vehicle engine parts.

The Doppler effect

When astronomers first began to look at the spectra of stars in other galaxies during the 1920s, they noticed that the spectra looked very similar to the spectra from stars in our own galaxy but that all the features present were shifted by the same relative amount towards the red end of the spectrum. This phenomenon became known as the **red shift**. This shift is due to the relative motion of the stars and our planet, in an effect called the **Doppler effect**. An observer receiving waves emitted from a moving body observes that the wavelength of the waves has been altered to a new wavelength.

A receding star or galaxy emits light that appears to have a *longer* wavelength than expected. We experience this effect when we hear a car coming towards us and driving past at a steady rate. As it approaches, the note of its engine rises to a maximum pitch, and then falls as the car travels away. You could imagine the waves getting squashed closer together (shorter wavelength) as the car drives towards you, and then stretched further apart (longer wavelength) as it drives away.

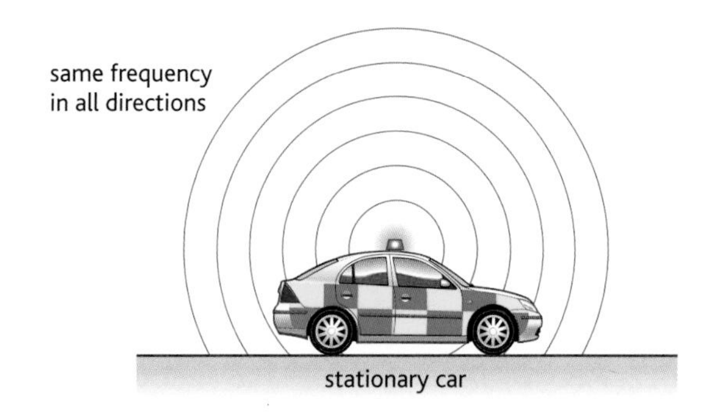

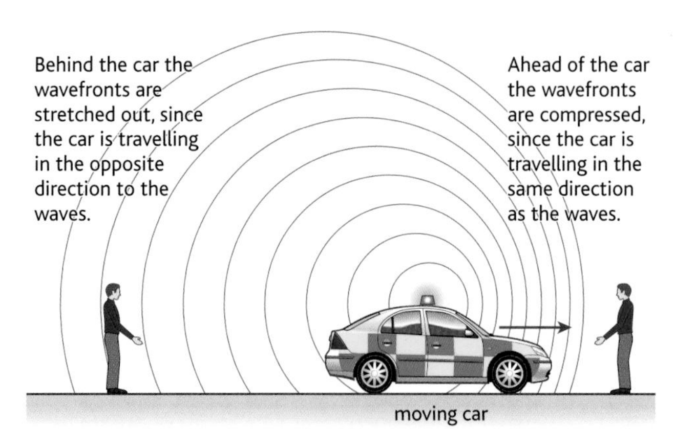

fig. 3.3.10 The Doppler effect causes a change in frequency and wavelength if there is relative motion between the wave source and the observer.

Astronomers quickly realised that the red shift implied that galaxies surrounding us were travelling away from us. In 1929 the American astronomer Edwin Hubble published his finding that the size of a galaxy's red shift is proportional to its distance from us – that is, the further away a galaxy is, the faster it is moving. Hubble's paper had the same effect on the twentieth century's view of the Universe as Galileo's work had some 300 years earlier. Instead of being static, the Universe was expanding. The philosophical implications continue to tax the minds of scientists and religious scholars.

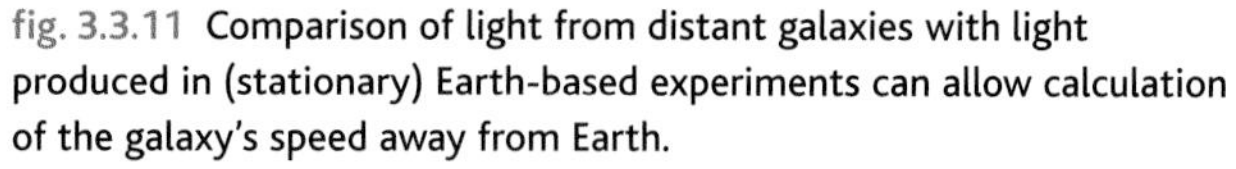

fig. 3.3.11 Comparison of light from distant galaxies with light produced in (stationary) Earth-based experiments can allow calculation of the galaxy's speed away from Earth.

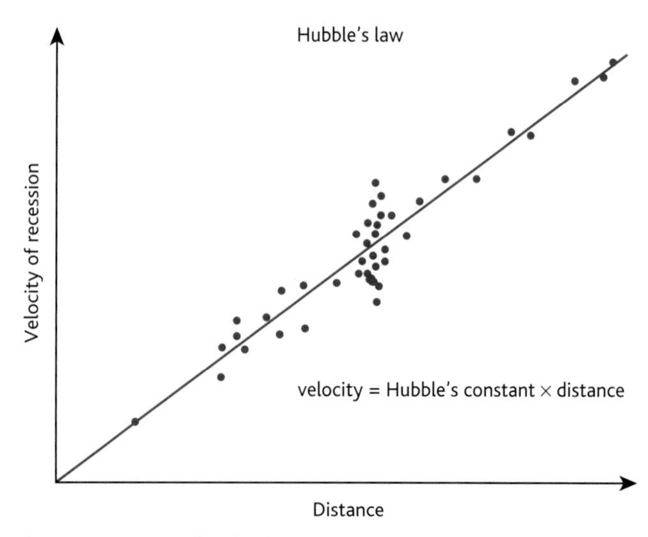

fig. 3.3.12 As virtually all galaxies show red shifts in their spectra, Hubble concluded that the all galaxies must be moving apart from each other and the Universe is expanding.

Pulse-echo detection

The natural habitat for bats tends to be woodland, and they fly through the trees at 10–15 $km\,h^{-1}$, depending on species. For catching insects, the bats need to be able to sense their location precisely. Whilst their vision is better than the old simile 'blind as a bat', their well-known echolocation system, using very high frequency (50–100 kHz) sound pulses, gives a much more detailed perception of the world at distances of less than 5 metres. At greater distances, the echo is attenuated too much for the bat to use. But how does it work? The bat will make a 'chirp' through its nose. This sound pulse will typically last 3 milliseconds. When the sound hits nearby objects, it will be reflected back to the bat's sensitive ears and its brain can accurately measure the time between pulse emission and echo reception. The bat's brain has also evolved to instinctively calculate the distance to the reflecting object using the equation:

$$\text{distance} = \text{speed} \times \text{time}$$

Experiments with dolphins have shown that they use a similar echolocation system, but their system is so sophisticated that dolphins can build up a mental image of the shapes of nearby objects. They can 'see' with sound. This is probably also true for bats, to ensure that they eat an insect and not an insect-sized leaf, but dolphins can respond to experimental inquiry in a more conclusive manner than bats.

We have developed similar pulse-echo ranging and imaging systems in a very wide range of technological applications, from sonar on ships and submarines to air traffic control radar, medical imaging and the measurement of distance to asteroids and to the Moon.

Worked example: Pulse-echo calculation

An air traffic control system sends out a pulse of radio waves that are reflected by a jumbo jet and the reflection is picked up by the radio dish 0.007 seconds after the emission of the pulse. How far away is the plane?

$$\text{speed} = \frac{\text{distance}}{\text{time}}$$

$$\text{distance} = \text{speed} \times \text{time}$$

$s = v \times t \quad v = \text{speed of light} = 3.0 \times 10^8\,m\,s^{-1}$

$s = 3.0 \times 10^8 \times 0.007 = 2.1 \times 10^6\,m = 2100\,km$

In that 0.007 seconds, the radio pulse has travelled to the plane and back again, so the actual distance to the plane is half that calculated:

$$\text{distance} = 0.5 \times 2100 = 1050\,km$$

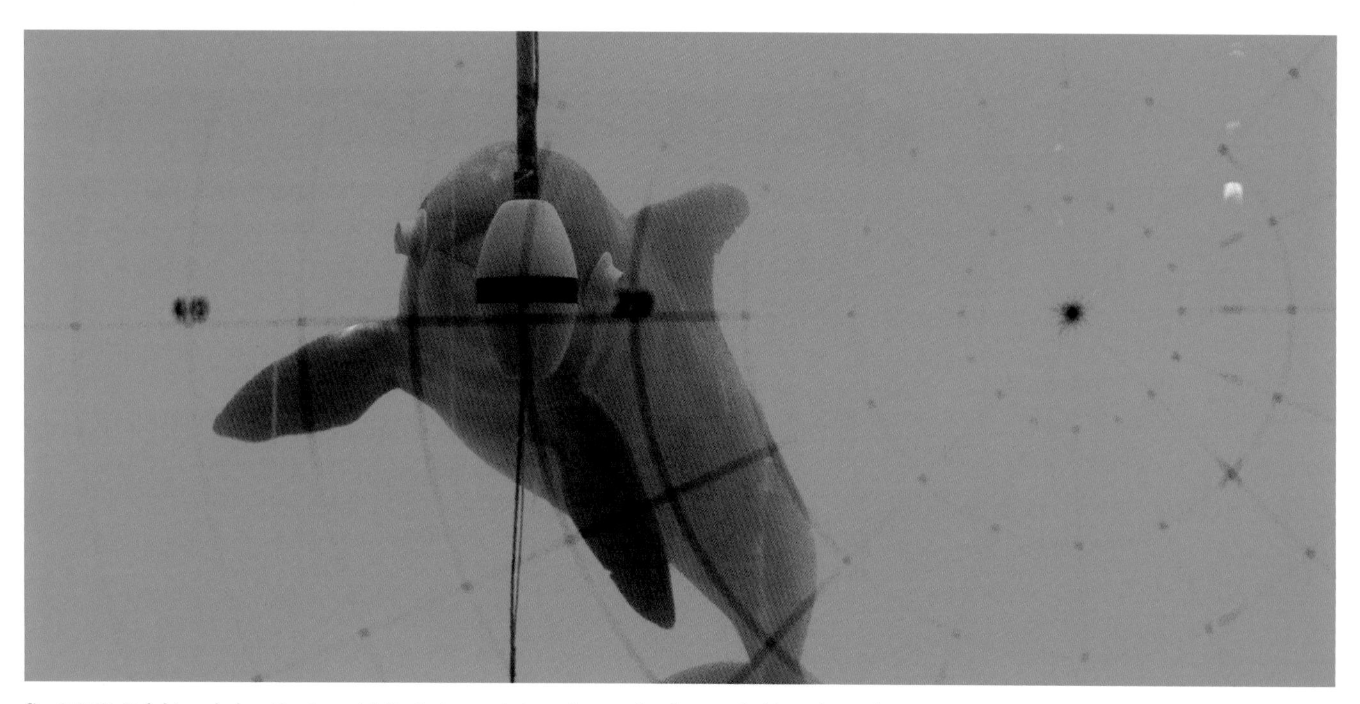

fig. 3.3.13 Dolphin echolocation is sophisticated enough to make out the shapes of objects in murky waters.

Doppler radar

Weather monitoring and air traffic control both use radar (**ra**dio **d**etection **a**nd **r**anging) to find the position and movement of storms and planes, respectively. In both cases, the system uses the basic pulse-echo idea. By measuring the time taken to reflect, and knowing the speed of the radio waves, a simple calculation will find the distance to the object. However, by also measuring any change in the wavelength – the Doppler shift – of the echo compared with the original pulse, the system can calculate the speed of the object reflecting the waves.

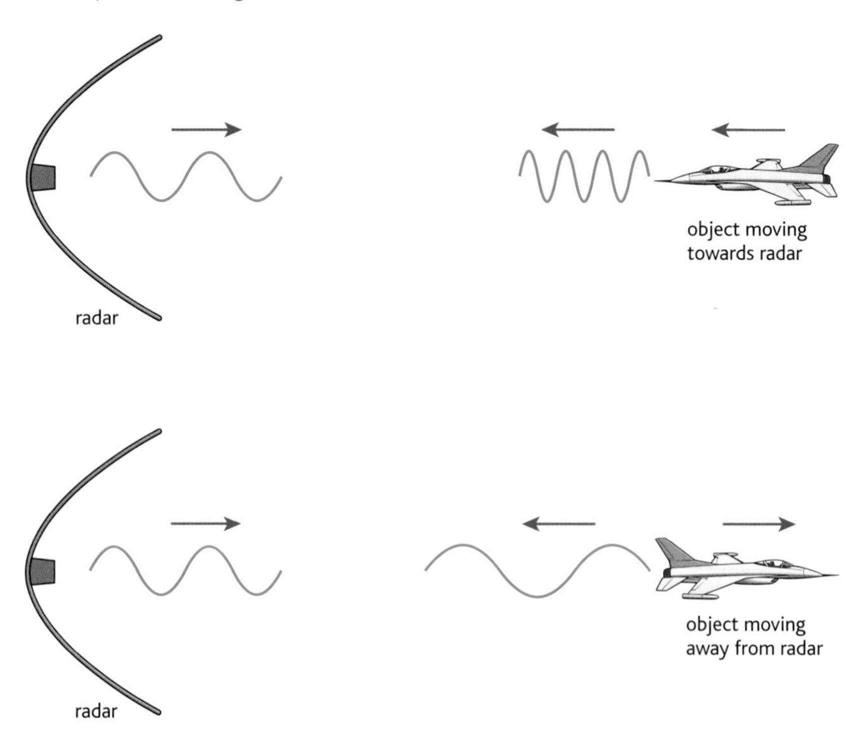

fig. 3.3.14 Radar systems use both pulse-echo calculations and Doppler shift calculations to monitor the position and velocity of objects.

Questions

1. Edwin Hubble discovered that the light from other galaxies is Doppler shifted towards the red end of the light spectrum. What does this tell us about the movement of galaxies? Why did this lead Hubble to conclude that the Universe is expanding?
2. **a** Why do spectators at a motor race hear the characteristic 'Neeeooww' sound as each car passes?

   b Why do the drivers hear a sound of constant pitch?
3. Why were Hubble's observations so important?

Ultrasound

Sound waves with frequencies from 20–20 000 Hz can be heard by human beings. This is our audible frequency range. Sounds with a frequency above 20 kHz are known as **ultrasound**. These waves are used in a variety of applications, in particular medical scans. With diagnostic scans, there is a huge medical advantage that an image of the insides of the patient can be produced without cutting them open. This is most familiar to people as scans on unborn babies.

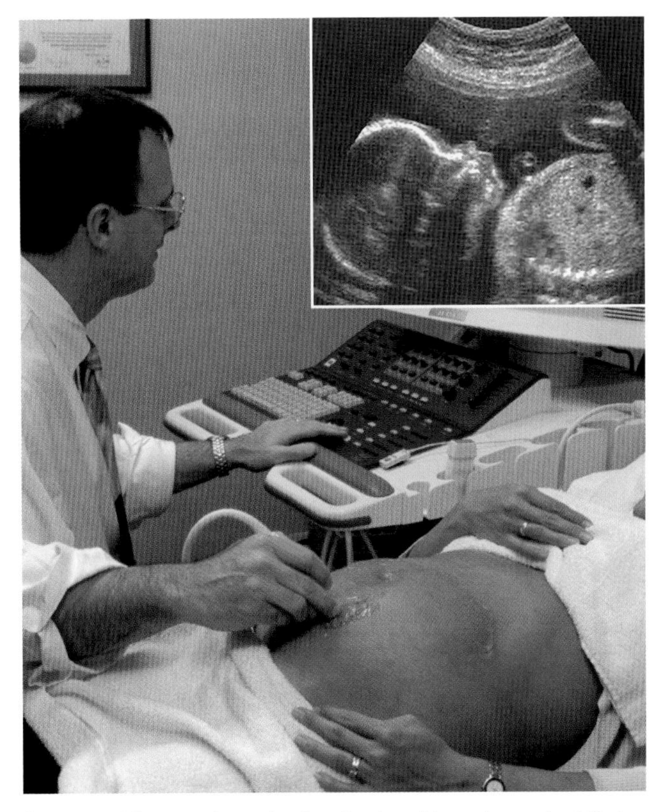

fig. 3.3.15 Ultrasound scanning has developed to such an extent that detailed moving images can now be produced with very user-friendly technology.

Ultrasound imaging relies on the fact that at any boundary between media, there will be a partial reflection of the sound waves. As a pulse of ultrasound waves passes from flesh to bone, there is a significant change in the density of the medium and the speed at which the waves travel. These changes cause a significant reflection of sound energy that can be detected. The type of tissue boundary involved in any reflection can be determined as this affects the proportion of energy reflected. The ultrasound transmitting device also acts as the detector for reflections, so it can accurately time how long the echo has taken to return, along with how much energy was reflected. With knowledge of the wave speed, a computer can calculate the depth within the body that each reflection occurs and build up an image showing features at different depths.

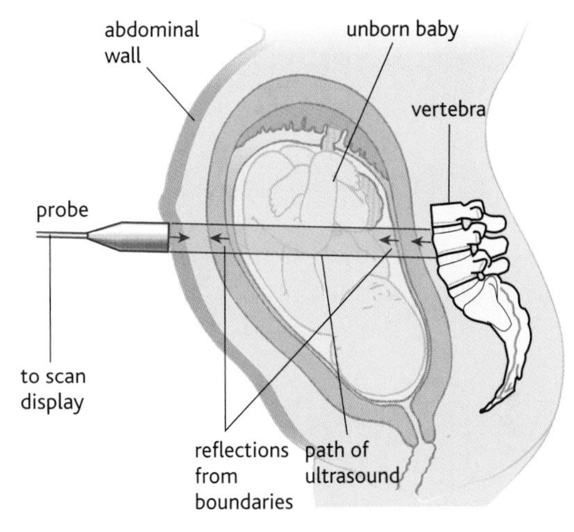

fig. 3.3.16 An ultrasound image is built up from detection of reflections at tissue boundaries within the body.

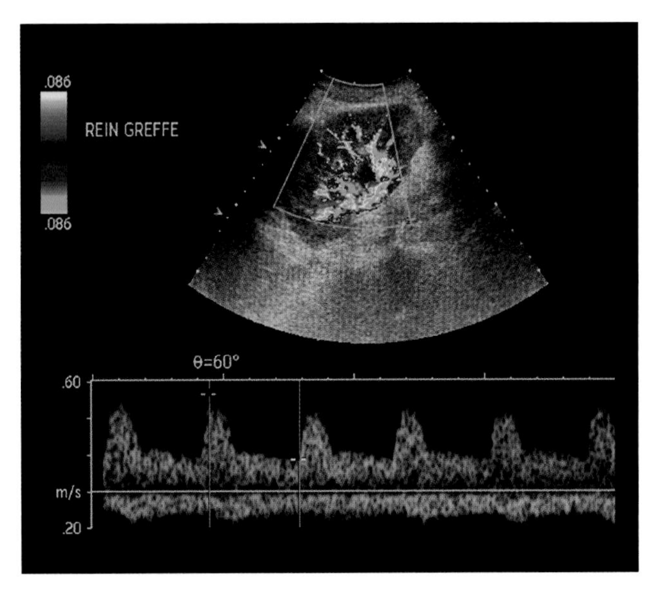

fig. 3.3.17 A Doppler ultrasound image.

The resolution of detail within an image is determined by the wavelength of the signal used. This means that the smaller the wavelength of sound used in an ultrasound scan, the smaller the finest detail that can be distinguished. There is a contrary trend though that the shorter the wavelength, the sooner a wave pulse will be absorbed. Medical ultrasound scans generally compromise these competing factors at a wavelength in the range 0.075 mm up to 1.5 mm, depending on the exact medical use. As a general rule of thumb in imaging, the smallest detail you can distinguish will be the same size as the wavelength. To make out a fetal thumb that is 0.5 mm wide would require the use of ultrasound with a wavelength of 0.5 mm or less. With pulse-echo imaging, there is an additional constraint that the resolution will be half the length of a pulse. A pulse may be a few wavelengths and its size can be calculated using:

length = speed × time pulse is on

Worked examples: Ultrasound resolution

An ultrasound system for examining the eye sends out a pulse of ultrasound waves with a frequency of 6 MHz. The pulse duration is 0.6□s. The speed of sound in the human eye averages $1510\,\mathrm{m\,s^{-1}}$. What is the resolution (smallest detail) of the image produced?

1 Pulse length method

distance = speed × time

$s = v \times t$, v = speed of sound = $1510\,\mathrm{m\,s^{-1}}$

pulse length $l = 1510 \times 0.6 \times 10^{-6} = 0.906 \times 10^{-3}\,\mathrm{m}$

resolution = half pulse length = 0.453 mm

2 Wavelength method

$v = f\lambda$

$$\lambda = \frac{v}{f}$$

$$\lambda = \frac{1510}{6 \times 10^{6}} = 2.51 \times 10^{-4}\,\mathrm{m} = 0.251\,\mathrm{mm}$$

From these two results, we take the larger answer that gives the worse resolution. Thus, the resolution is 0.453 mm and finer details than this could not be seen in the image produced.

Doppler ultrasound

If the ultrasound pulse in a medical scan were to encounter a moving reflector, flowing blood for example, it would be reflected so the position could be determined. However, there would be a Doppler shift in the frequency of the sound waves being used. If the Doppler shift is measured, then the speed of movement of the blood could be calculated. A 3 MHz ultrasound beam travelling at $1500\,\mathrm{m\,s^{-1}}$ would be shifted by about 600 Hz if the blood were moving at $0.15\,\mathrm{m\,s^{-1}}$. This method is used by modern ultrasound imaging machines to measure speeds of movements of many internal parts of the body see **fig. 3.3.17**.

HSW How safe is antenatal scanning?

Ultrasound imaging is used worldwide as a diagnostic tool in medicine. In particular, antenatal scanning of babies is now a routine part of pregnancy in many countries. Antenatal scanning using X-rays is not permitted because these waves are known to be a cause of cancer with extensive exposure. Might the same not be true of ultrasound?

Ultrasound imaging has developed widely, growing out of therapeutic uses in which the sound waves are used to disrupt tissue that is suffering injury or disease, and combining these with additional input from the technology used to detect flaws in metal. The development to today's procedures took approximately 50 years, with the first steps being made during the Second World War. The technology developed separately in different countries in a somewhat arbitrary trial-and-error fashion. Very few scientific studies were done to discover the long-term effects of ultrasound exposure. Only recently have studies shown some bad effects being more likely with regular exposure during pregnancy, such as lower birthweight babies, premature births and infants taking longer to develop speech. The UK Government's Health Protection Agency (HPA) disputes the statistical significance of these studies, and further highlights the scientific difficulties in determining the degree of danger. These problems include:

- very few children are not scanned, and thus it is difficult to find a control group for which all other lifestyle factors are constant
- future monitoring of scanned children will not be relevant as manufacturers constantly change the sound wave characteristics in use in order to improve image quality
- the extent to which studies on mice or monkeys are applicable to humans.

Might it turn out that, in a generation, we will find all those exposed to ultrasound whilst in the womb dying young from 'ultrasonic cancer'? The process is so ingrained within society that it may now prove difficult to convince people that it could be more harmful than beneficial. As there are known biological effects at high powers, or after long exposures, the HPA recommend against any clinically unnecessary exposure, and are particularly concerned with the growth industry that offers 'souvenir' videos of your unborn baby.

Questions

1 In 1959, the distance to the Moon was established by the US Navy using radar that gave the distance as being 384 400 km, with an uncertainty of 1.2 km in this figure. Explain the basics of the pulse-echo technique that would have been used for this measurement.

2 Give one similarity and one difference between air traffic control radar and a bat's echolocation system.

3 A weather station monitors a storm using a radar system. The radio wave pulse emitted is received back at the transmitter after 0.34 ms. The wavelength emitted was 5.6 cm and that received after reflection from the storm was 1 billionth of a metre longer. What information about the storm can be worked out from these data?

4 An ultrasound scanner for unborn babies has the following specifications printed on it:

Frequency = 3 MHz

Pulse duration = 1□s

If the speed of sound in the womb is $1520\,\mathrm{m\,s^{-1}}$, what would be the smallest detail that could be distinguished?

Examzone: Topic 3 Waves

1 The diagram below shows a loudspeaker which sends a note of constant frequency towards a vertical metal sheet. As the microphone is moved between the loudspeaker and the metal sheet the amplitude of the vertical trace on the oscilloscope continually changes several times between maximum and minimum values. This shows that a stationary wave has been set up in the space between the loudspeaker and the metal sheet.

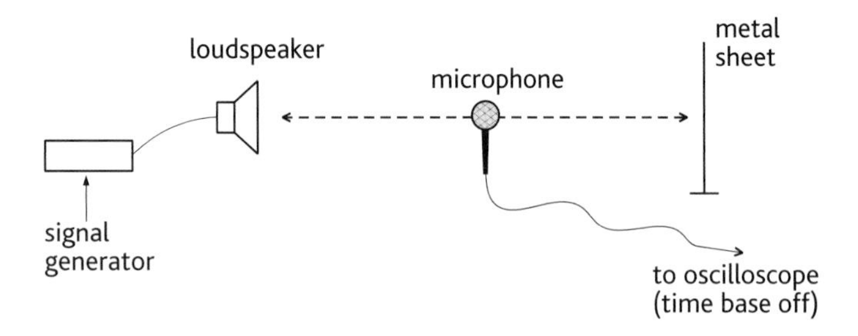

How has the stationary wave been produced? **(2)**

Calculate the speed with which the stone leaves the catapult. **(2)**

What measurements would you take, and how would you use them, to calculate the speed of sound in air? **(4)**

Suggest why the minima detected near the sheet are much smaller than those detected near the loudspeaker. **(2)**

(Total 10 marks)

2 The diagram shows the shape of a wave on a stretched rope at one instant of time. The wave is travelling to the right.

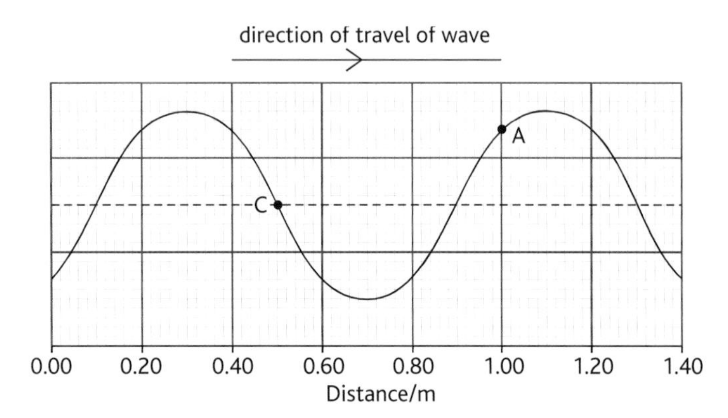

Determine the wavelength of the wave.

Copy the diagram and mark the point on the rope whose motion is exactly out of phase with the motion at point A. Label this point X.

Mark on your diagram a point on the rope which is at rest at the instant shown. Label this point Y.

Draw an arrow on your diagram at point C to show the direction in which the rope at C is moving at the instant shown. **(4)**

The wave speed is $3.2\,\mathrm{m\,s^{-1}}$. After how long will the rope next appear exactly the same as in the diagram above? **(2)**

(Total 6 marks)

3 A microwave generator produces plane polarised electromagnetic waves of wavelength 29 mm.

a (i) Calculate the frequency of this radiation. **(1)**

(ii) Name the missing parts of the electromagnetic spectrum labelled a–e.

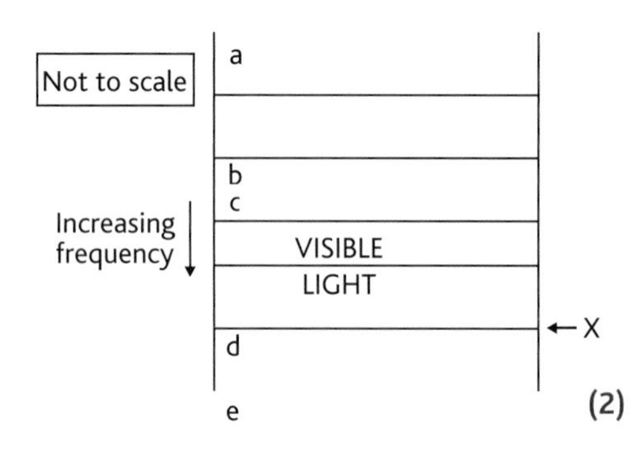

(2)

(iii) State a typical value for the wavelength of radiation at boundary X. **(1)**

b Describe, with the aid of a diagram, how you would demonstrate that these microwaves were plane polarised. **(4)**

(Total 8 marks)

4 Copy and complete the diagram below to show the different regions of the electromagnetic spectrum. **(2)**

Radio waves	

State four differences between radio waves and sound waves. **(4)**

Two radio stations broadcast at frequencies of 198 kHz and 95.8 MHz. Which station broadcasts at the longer wavelength? **(1)**

Why do obstacles such as buildings and hills present less of a problem for the reception of the signal from the station transmitting at the longer wavelength? **(2)**

(Total 9 marks)

Topic 4 DC electricity

Electricity is integral to modern everyday life. In the developed world, we take electrical devices for granted. Rarely will you consider the flow of electrons through an MP3 player, or the electrical energy needed to run a video game.

This topic covers the definitions of various electrical quantities, for example current and resistance, Ohm's law and non-ohmic conductors, potential dividers, emf and internal resistance.

What are the theories?

You may already be familiar with Ohm's law for calculating resistance, but you will also learn how to calculate it from a knowledge of resistivity, and how it affects the potential difference across different components in a circuit.

Kirchhoff's laws, which apply conservation of charge and energy to electric circuits, allow calculation of current and voltage in different parts of circuits.

If we consider electrical conduction on a subatomic scale, we can calculate how many electrons move and how fast they move. We can also understand why some materials are insulators or semiconductors, and why the resistance changes with temperature.

What is the evidence?

There are many opportunities in this topic for you to confirm, or independently discover, the rules regarding electric circuits that scientists currently believe. These range from something as simple as calculating the resistance having measured current and voltage, through plotting graphs to show how current varies with voltage, which can be a check on changes in resistance, up to measurements of the internal resistance of power supplies or their power output.

These days, traditional needle meters are likely to have been replaced in your school laboratory by digital meters. But the march of progress doesn't end there – you may also have an opportunity to see how current and voltage vary over time by taking measurements using sensors connected to a datalogging computer.

What are the implications?

The theories that you will learn in this topic are really only valuable in their use and application. From adjusting the volume on an iPod to checking the power produced by an industrial generator, the ability to calculate quantities in electric circuits is an incredibly important capability.

The map opposite shows you all the knowledge and skills you need to have by the end of this topic. The colour in each box shows which chapter they are covered in and the numbers refer to the sections in the Edexcel specification.

Chapter 4.1

describe electric current as the rate of flow of charged particles and use the expression $I = \Delta Q/\Delta t$ (50)

interpret current–potential difference graphs, including non-ohmic materials (56)

demonstrate an understanding of how ICT may be used to obtain current–potential difference graphs, including non-ohmic materials, and compare this with traditional techniques in terms of reliability and validity of data (55)

define the concept of emf (part of 59)

use the fact that resistance is defined by $R = V/I$ and that Ohm's law is a special case when $I \propto V$ (54)

use the expression $V = W/Q$ (51)

investigate and use the relationship $= \rho l/A$ (57)

use $I = nqvA$ to explain the large range of resistivities of different materials (61)

Chapter 4.2

investigate and use the expressions $P = VI$, $W = VIt$. Recognise and use related expressions, for example $P = I^2R$ and $P = V^2/R$ (53)

investigate and explain how the potential difference along a uniform current-carrying wire varies with distance along it and how this variation can be made use of in a potential divider (58)

explain, qualitatively, how changes of resistance with temperature may be modelled in terms of lattice vibrations and number of conduction electrons (62)

define internal resistance. Use the concepts of emf and internal resistance and distinguish between emf and terminal potential difference (remainder of 59)

recognise, investigate and use the relationships between current, voltage and resistance, for series and parallel circuits, and know that these relationships are a consequence of the conservation of charge and energy (52)

investigate and recall that the resistance of metallic conductors increases with increasing temperature and that the resistance of negative temperature coefficient thermistors decreases with increasing temperature (60)

4.1 Electrical quantities

Introducing electricity

Those of us who live in the developed world take electricity for granted. Its usefulness lies in its ability to provide the energy to run devices ranging from underground trains to digital watches, and the ease with which that energy may be transmitted through wires and cables. Yet for all this versatility, it is important to remember that electricity is only a means of transferring energy, not a source of energy in its own right (in the language of the energy business it is a **secondary source**, not a **primary source** like oil, coal or gas). Electricity therefore has to be made, either by mechanical means using generators or by using chemicals in cells and batteries, or even by direct conversion of the energy in sunlight into electricity.

A brief history of electricity

Electricity has been known since ancient times, when it was discovered that a rubbed piece of amber would attract small hairs and other light objects – the Greek word for amber (*ηλεκτρογ*) gives us our modern word **electron**. The idea that electricity involved something called **charge** of two types, positive and negative, was established by the work of Du Fay in the eighteenth century. However, the exact meaning of this was unclear to physicists for over 150 years, needing the development of theories about atomic structure for an explanation. Both Du Fay and Benjamin Franklin (famous for his experiment in which he flew a kite in a thunderstorm to show that lightning is a form of electricity) believed that electricity was a kind of fluid. Whilst we know now that this model is not a good one, it is as well to remember that the flow of electric current can be modelled quite effectively using the same mathematical relationships as those used to model the flow of a fluid through a pipe.

Our present model of electric current as a flow of charged particles took many years to develop, and was not possible to establish fully until work at the beginning of the last century established the structure of the atom. **Fig. 4.1.1** shows how some of our ideas about electricity were established up to the time of Faraday in the middle of the nineteenth century – further examination of the discovery of charged particles and their effects can be found in Topic 5.

The fact that amber could be made to attract hairs, dust and other small objects by rubbing it is known to the Ancient Greeks. Our modern explanation of this is in terms of charge, an idea developed in the eighteenth century.

Charles François de Cisternay Du Fay
In 1733 Du Fay published his discovery that there are two types of charge, and that like charges repel while unlike charges attract.

Pieter van Musschenbroek
Van Musschenbroek received the first known electric shock (other than those caused by lightning) in 1747, having invented the Leiden jar, a device for storing electric charge. He reported, 'in a word, I thought it was all up with me' after his experience.

Benjamin Franklin
Van Musschenbroek's work inspired Franklin to carry out an experiment in 1752 in which he flew a kite during a thunderstorm to show that lightning is a kind of electricity similar to that stored by the Leiden jar. Many other experimenters who tried this experiment were killed.

Joseph Priestley
Franklin encouraged Priestley to write *The history and present state of electricity*, an account of his work on electricity published in 1767. In it, Priestley suggested that the forces between electrical forces follow an inverse square law. Detailed experimental observations confirming this are carried out by Charles-Augustin Coulomb in France in 1785 – the inverse square relationship for electric charges becomes known as Coulomb's law.

Luigi Galvani
In 1791 Galvani announced the discovery of 'animal electricity', having observed that the muscles in severed frogs' legs twitch when touched by two different metals.

Alessandro Volta
In 1794 Volta showed that the effect of the two metals described by Galvani was unconnected with living things, and that electricity can be produced whenever two metals are immersed in a conducting solution.

Hans Christian Oersted
Oersted accidently placed a compass near a wire carrying an electric current while performing a classroom demonstration in 1820. He noticed that the compass needle moved and concluded that electricity and magnetism are linked.

André-Marie Ampère
As a result of Oersted's work, Ampère investigated the force between two current-carrying conductors and concluded that the wires repel or attract each other depending on the relative directions of the currents.

Michael Faraday and Joseph Henry
During the 1820s and 1830s, Henry and Faraday investigated the relationship between electricity and magnetism independently, Faraday in England, Henry in America. Both men reached similar conclusions and made similar discoveries (they both discovered the principle of the dynamo at about the same time, for example). While Faraday's discoveries were to have a major influence on the theoretical basis of physics, Henry's work resulted in immediate practical applications like the telegraph, and ultimately the telephone.

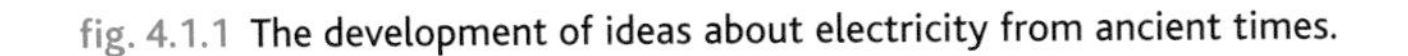

fig. 4.1.1 The development of ideas about electricity from ancient times.

Electric current

Flow of charge

The movement of charged particles that we call an electric current is something that cannot usually be seen – for example, we cannot observe electrons flowing through a metal wire. We can see the effect of an electric current however, which may cause:

- a magnetic field
- chemical changes (for example, the formation of copper metal when a current flows through a solution containing copper ions)
- an increase in temperature of whatever is carrying the current.

If we have a wire carrying a current I, then the total amount of charge ΔQ passing a given point in time Δt will depend on I and t. This means that

$$\Delta Q = I\Delta t$$

The SI unit of charge is the **coulomb** (C), where 1 C = 1 A s, that is, the charge passing a given point in a wire when a current of 1 A flows through the wire for 1 s. We can also write:

$$I = \frac{\Delta Q}{\Delta t}$$

which shows that current is the rate of flow of charge.

The unit of current, the **ampere** (A), could also be defined as the current when one coulomb of charge flows each second. One coulomb is the amount of charge on 6.25×10^{18} electrons (although strictly speaking with electrons this should be quoted as a negative value: $Q = -1$ C).

Worked example

If 12.5×10^{18} electrons pass through a lamp in half a second, what is the current flowing?

From the definition of current:

$$I = \Delta Q / \Delta t$$

Here ΔQ in coulombs is given by:

$$\Delta Q = \frac{12.5 \times 10^{18}}{6.25 \times 10^{18}} = 2\,\text{C}$$

We have 2 coulombs of charge, so:

$$I = \frac{2}{0.5} = 4\,\text{A}$$

Graphing current, charge and time

If we are dealing with a changing current, we could draw a graph showing the current over time.

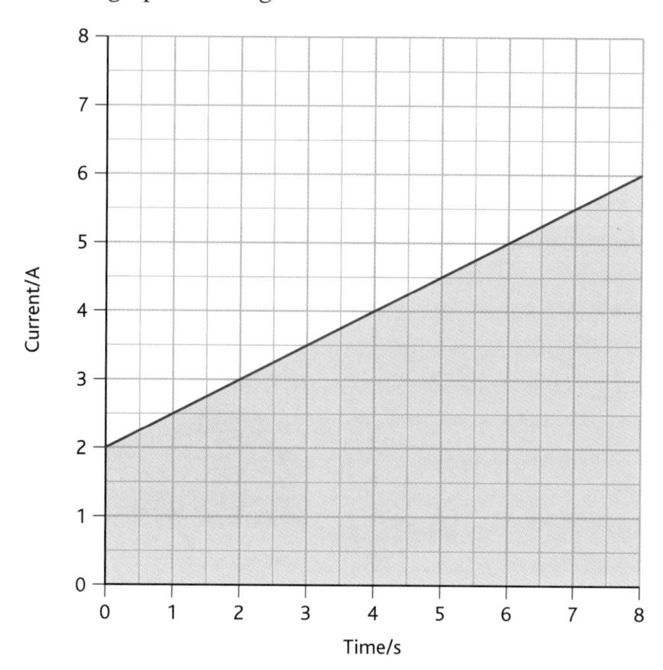

fig. 4.1.2 If we find the area under the line on this graph, it comes to 32. This means that 32 coulombs of charge passed in the 8 seconds shown.

Measuring the area under the graph is equivalent to multiplying current by time at each moment and adding up all the answers. This means it is a way to find the total charge that passes in that time period.

Circuits

For an electric current to flow a complete circuit is needed. To explain this, a model of electricity is often used in which water flowing through pipes represents electric current. Although this is a satisfactory model in some respects it is unsatisfactory in others, and it is as well to recognise this now – like any model, the 'electricity as a fluid' model collapses if pushed too far.

Consider a water circuit in which water causes a turbine to rotate (**fig. 4.1.3a**).

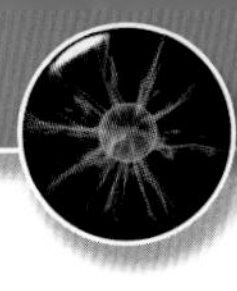

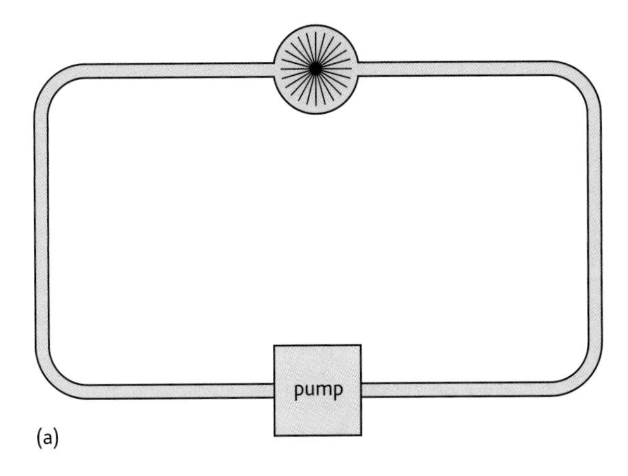

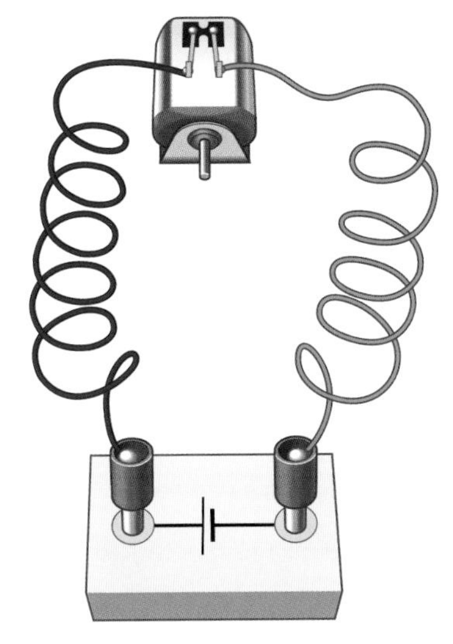

(b)

fig. 4.1.3 **a** Turbine driven by water **b** Electric motor driven by a battery

This situation is similar to a battery driving an electric motor, shown in **fig. 4.1.3b**. In each case 'something' travels round a complete circuit. Energy is transferred to the system at one point and transferred from it at another, and appears to be carried from one place in the circuit to another by the 'something' that moves round the circuit. Studies of the structure of metallic conductors suggest that it is free electrons in metals that make up the current in an electrical circuit – an electric current in a metal is nothing more than a flow of electrons carrying negative charge. Electric current is not always a flow of electrons, however. For example, in an electrolyte (a conducting solution – salt solution for instance) positive and negative ions act as charge carriers. By convention, electric current flows in the direction of the flow of positive charge, so electrons flow one way round a circuit while conventional electric current flows in the other (**fig. 4.1.4**).

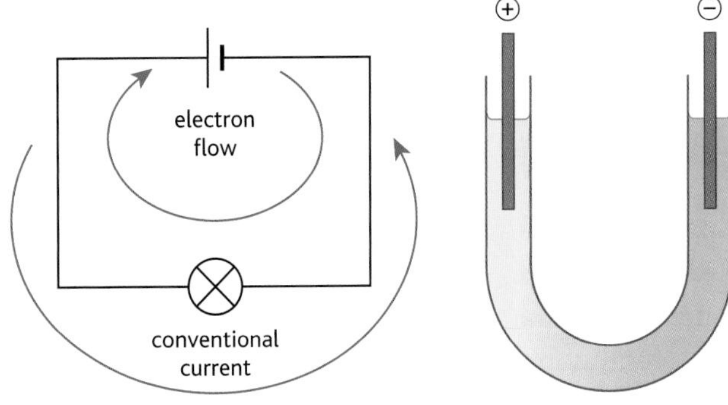

fig. 4.1.4 Conventional current and the flow of electrons in a circuit.

It does not matter whether you use the direction of flow of electrons or the direction of flow of conventional electric current when referring to the current through a wire – but you must be consistent. In this book we shall use the flow of conventional current in all cases. **Fig. 4.1.4** shows current flowing through a solution containing blue copper ions and yellow chromate ions. The positive copper ions move in one direction, while the negative chromate ions move in the opposite direction – in this case both are involved in the flow of electric current.

Questions

1 A current of 0.2 amperes flows through a light bulb whilst it is on for 20 minutes. How much charge flows through the bulb in that time?

2 6.25×10^{18} electrons together have a total negative charge of 1 coulomb. How much charge does each electron carry?

3 **a** Draw a circuit diagram to show a bulb and a motor in series with a battery of three cells. Include an ammeter to measure the current through the motor.

b **i** Add a labelled arrow to your circuit showing the direction of conventional current in the circuit.

ii Add another labelled arrow showing the direction of electron movement in the circuit.

Energy and electricity

Electromotive force

Anyone who has carried out simple investigations into electrical circuits is familiar with the need for an energy source to make an electric current flow. In many cases this energy source is a cell, in which chemicals react to cause an electric current to flow when the terminals of the cell are joined by a conductor. The energy changes occurring in this situation are sometimes modelled in the way shown in **fig. 4.1.5**.

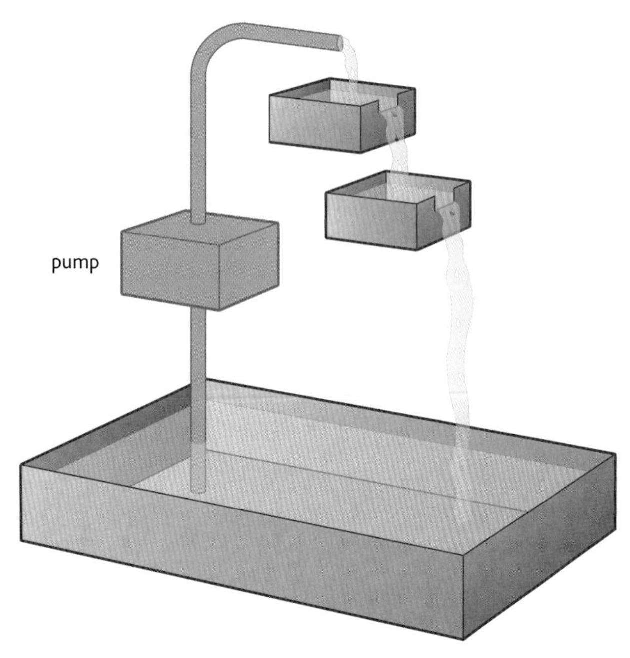

fig. 4.1.5 In this model of an electric circuit the cell is modelled as a device that increases the gravitational potential energy of water in the circuit, which then flows round the rest of the circuit under the influence of gravity.

The 'cell' in the water circuit in **fig. 4.1.5** provides energy to each bucket of water that passes through it, and the amount of energy per bucket is a measure of the effect of the cell on the flow of water round the circuit – the more energy per bucket the greater the flow of water. This is true for the electrical circuit too, where we can measure the amount of energy supplied by the cell to each 'bucket' of charge that flows through it. The amount of energy supplied to each unit of charge is called the **electromotive force** or **emf** of the cell. In the SI system, the unit of emf is the **volt** (V). A cell is said to have an emf of 1 V when it supplies 1 J of energy to each 1 C of charge flowing through it, in other words:

$$1\,\text{V} = 1\,\text{J}\,\text{C}^{-1}$$

Of course there are other energy changes taking place in a circuit too. Just as the energy stored in the chemicals of a cell can cause charges to flow through a circuit, the energy carried by moving charges can cause (for example) the filament of a lamp to become white hot and emit heat and light. In this case we refer to a **potential difference** or **pd** across the object where the energy carried by the current causes change to occur. Potential difference has the same units as emf, since it measures energy transferred per unit charge. In this case, however, the energy is transferred from the charge to the lamp. This means that pd is a measure of work being done (remember that work is defined as the amount of energy transferred).

This gives us the definition of potential difference, V, as work done, W (or energy transferred) per unit of charge, Q:

$$V = \frac{W}{Q}$$

Worked example

If 25×10^{18} electrons pass through a lamp and it transfers 48 joules of electrical energy into heat and light, what is the potential difference of the lamp?

Definition of pd: $V = \frac{W}{Q}$

Here we have 4 coulombs of charge, so:

$$V = \frac{48}{4} = 12\,\text{V}$$

Questions

1 a Write a definition in words, and also as an equation, for 'electromotive force'.

b Explain how pd is different from emf.

2 a A cell provides 88 C of charge with 132 J of energy. What is the emf of the cell?

b When a 3A current lights a light bulb, it receives 120 J of energy in 5 s. What is the potential difference across the light bulb?

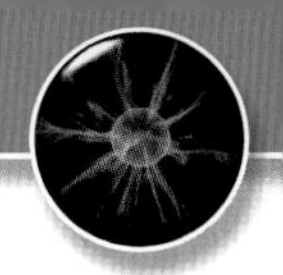

HSW The flow model for electricity

Thinking of electric current too literally in terms of flow can be unhelpful. One example will illustrate the problem.

A lamp may be connected to a suitable cell via a switch using leads that are (say) 2 m long. If the switch is closed, the lamp lights almost immediately. One explanation for this is to say that energy is being transferred from the cell to the lamp, and that this energy is being carried by a flow of electrons. This is very similar to the way in which water in the heating circuit of a central heating system carries energy from the boiler to the radiators – a very straightforward analogy.

The situation seems less straightforward if we look at the speed at which the electrons travel through the wire. We cannot do this by experiment, but some fairly simple physics leads us to a relationship – the **transport equation** – which enables us to calculate this. In the situation we have described above, with copper wires connecting the cell and lamp together, the electrons in the copper are unlikely to be travelling faster than one hundredth of a millimetre in each second. This suggests that an electron would take around 2 days to travel from the cell to the lamp – which makes it seem unlikely that it is the flow of electrons that is involved in transferring energy from the cell to the lamp. A practical investigation to measure the speed of an electric pulse along a wire confirms that the speed is indeed very large – so large that it is close to that of light. This suggests that the energy may be carried by an electromagnetic wave or something similar to it, and that a simple visual model of electrons 'picking up' energy as they pass through the cell and then 'unloading' it as they pass through the lamp is totally inadequate. Instead, we should visualise the cell as causing an electric field in the conductor. It is this field that causes the current to flow, and that provides the means for transferring the energy from cell to lamp.

The word **voltage** is often used in situations where the more precise terms emf or pd should really be used. This is not always important, but the term voltage does not tell you whether energy is transferred *to* or *from* the electric charge flowing round the circuit, whereas the terms emf and pd do. Since current flows *through* a conductor, it makes sense to say things like 'the current through the lamp is 0.2 A'. Similarly, pd is measured *across* components in a circuit, so it is sensible to make statements like 'the pd across the motor is 2.5 V'. It is particularly important not to talk in terms of 'the voltage *through* the lamp' or whatever – such statements are nonsense! However, emf is usually thought of as a property of a cell or battery in a circuit, so it is reasonable to speak of 'the emf of the battery'.

Measuring current and potential difference

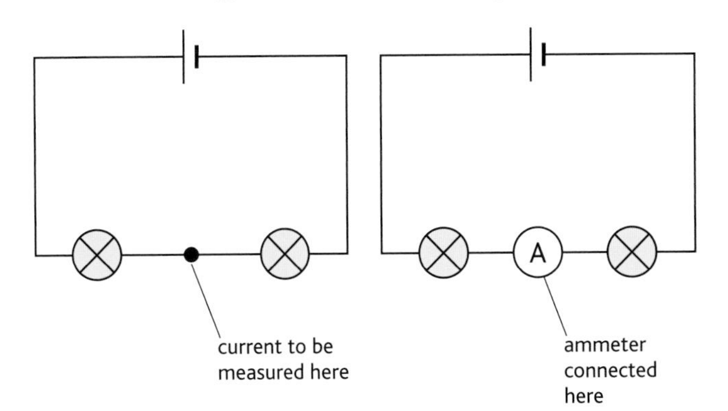

fig. 4.1.6 Ammeters are used to measure the current at a specific point in a circuit. They are therefore connected into the circuit in series, at the point where the current is to be measured.

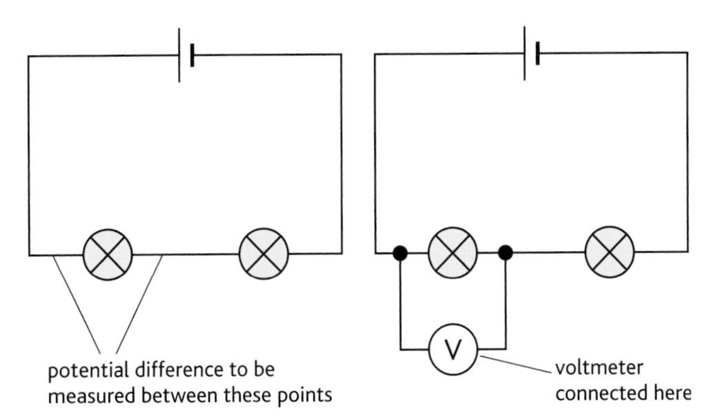

fig. 4.1.7 Voltmeters are used to measure the potential difference between two points in a circuit. They are therefore connected into the circuit in parallel, between the points where the potential difference is to be measured.

Nowadays, current and voltage sensors are often connected to a computer that can record the current and pd over time and record many more readings than a human, with fewer errors, and at much more frequent intervals. The software can then be used to produce graphs of the readings (such as that in **fig. 4.1.9**) showing how the current varies with the pd across a component, quickly and easily. The use of computers to take and record measurements improves the reliability and validity of the data in such experiments.

Resisting current flow

Resistance

The causes of electrical resistance will be discussed later. However, its value in any component can be calculated at any given moment if we know the pd and current at that instant. Resistance is defined mathematically as:

$$R = \frac{V}{I}$$

Worked example

If a 12 volt lamp carries a current of 0.5 amperes, what is the resistance of the lamp?

Resistance:

$$R = \frac{V}{I} = \frac{12}{0.5} = 24\ \Omega$$

The relationship between pd and current

Although current and pd measure different things, they are related to each other. If we connect a battery up to a length of resistance wire, we can measure the current through the wire if we use 1, 2, 3 or 4 cells (**fig. 4.1.8**). The results of such an investigation are shown in **table 4.1.1** and plotted as a graph in **fig. 4.1.9**.

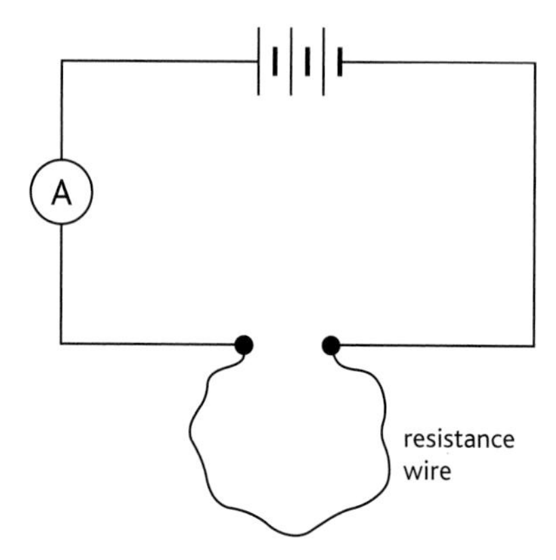

fig. 4.1.8

Number of cells	Potential difference across ends of wire/V	Current/A
1	1.5	0.2
2	3.0	0.4
3	4.5	0.6
4	6.0	0.8

table 4.1.1

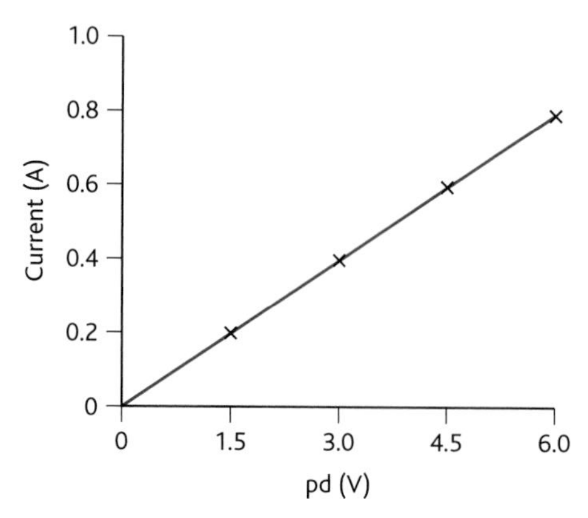

fig. 4.1.9 Graph of the results shown in table 4.1.1.

The results show that there is a simple relationship between the pd across the ends of the wire and the current through it – doubling the pd doubles the current. Repeating the investigation with other wires gives results that show a similar relationship between pd and current. The graphs are always straight lines through the origin, although the slopes are different. This relationship was first discovered by the German physicist Georg Ohm in 1826 and is called Ohm's law:

> **Provided the temperature and other physical factors remain constant, the current through a wire is proportional to the potential difference across its ends.**

If we divide the potential difference across the ends of the wire by the current through it, we get a constant figure for a given piece of wire. The figure is the resistance of the wire, and represents its opposition to current. A conductor has a resistance of 1 ohm if a current of 1 A flows through it when it has a pd of 1 V across it. (1 ohm is sometimes written as 1 Ω.) This leads to a very simple mathematical relationship which states Ohm's law:

$$I = \frac{V}{R}$$

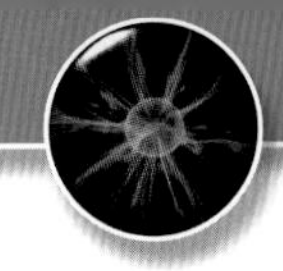

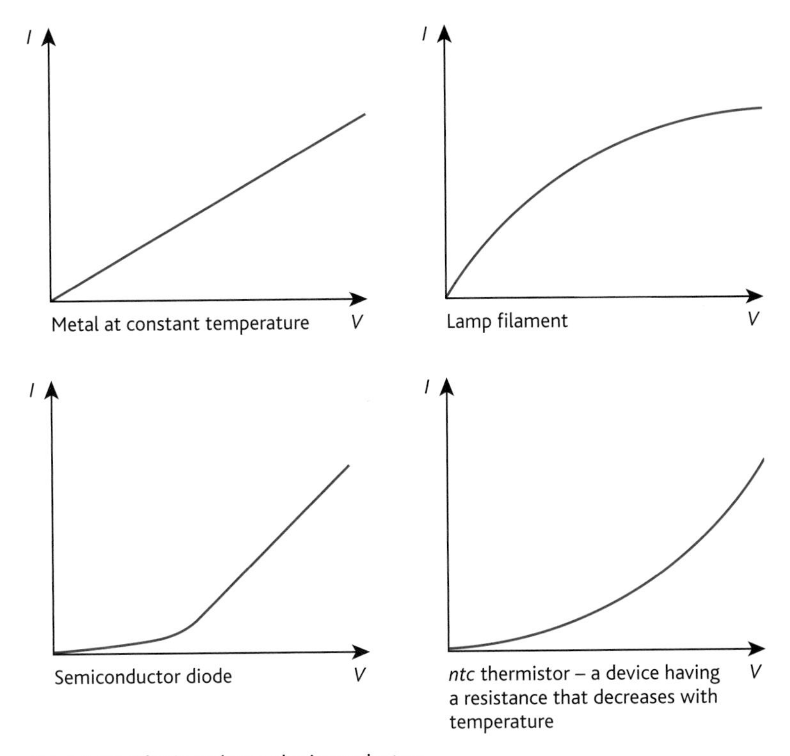

fig. 4.1.10 Ohmic and non-ohmic conductors.

Ohmic and non-ohmic conductors

Not all circuit elements have the simple current–voltage graphs seen in fig. 4.1.9. If the relationship between the current through a conductor and the pd across its ends is explored, the relationship is found to vary according to the circuit element. Conductors for which I–V graphs are straight lines are said to be **ohmic conductors** – a metal at a constant temperature is an example of such a conductor. Where the I–V graph is non-linear, the conductor is said to be **non-ohmic** – a lamp filament and a semiconductor diode are two examples of non-ohmic conductors. At any point on the I–V line, the gradient of the line will represent current divided by pd, so it will be the inverse of the resistance at that point. This allows us to see how the resistance of a component changes with the current through it (which can indicate the temperature of the component). The change in resistivity with temperature is explored further in the section 'Understanding conduction'.

The resistance of a conductor

As the simple investigation described above shows, the resistance of two different wires is not necessarily the same. Investigations show that the resistance of a uniform conductor depends on:

- its length (l)
- its cross-sectional area (A)
- the material of which it is made.

The relationship between resistance and these three quantities is given by:

$$R = \frac{\rho l}{A}$$

where ρ is the **resistivity** of the material from which the conductor is made. In the SI system resistivity is measured in Ωm.

Material	Resistivity ρ/Ωm at 20 °C	$\Delta\rho / \Delta t$ / % C^{-1}
Silver	1.6×10^{-8}	+0.38
Copper	1.7×10^{-8}	+0.39
Aluminium	2.8×10^{-8}	+0.2
Constantan	4.9×10^{-7}	+0.001
Germanium	4.2×10^{-1}	
Silicon	2.6×10^{3}	
Poly(ethene)	2×10^{11}	
Glass	$\sim 10^{12}$	
Epoxy resin	$\sim 10^{15}$	

table 4.1.2 Resistivity varies greatly between materials, and is also dependent on temperature. Note the small change in resistivity with temperature for constantan, an alloy of copper and nickel, used where accurately known resistance is important. As some materials are cooled to very low temperatures, superconductivity may result.

Worked example

What resistance would there be through a 5 cm cube of constantan metal?

$l = 0.05$ m

$A = 0.05 \times 0.05 = 2.5 \times 10^{-3}$ m^2

From table 4.1.2, $\rho = 4.9 \times 10^{-7}\,\Omega$m, so:

$$R = \frac{\rho l}{A}$$

$$R = \frac{4.9 \times 10^{-7} \times 0.05}{2.5 \times 10^{-3}}$$

$$R = 9.8 \times 10^{-6}\,\Omega$$

Superconductivity

The Dutch physicist Heike Kamerlingh Onnes discovered how to liquefy helium (which has a boiling point of just 4.2 K) in 1908, and opened the door to the study of low-temperature physics. Only 3 years later, Onnes discovered that mercury cooled to 4.15 K completely loses its resistivity and becomes a perfect conductor – a **superconductor**.

A superconductor is a perfect conductor – a closed loop of superconducting wire carrying a current will continue to carry the current as long as it is kept below the critical temperature T_c below which it becomes a superconductor. Such a current is called a **persistent current**.

fig. 4.1.13 J. Georg Bednorz and K. Alex Miller.

Several other metals apart from mercury exhibit superconductivity, as do many compounds and alloys. Some of the compounds recently discovered become superconductors at relatively high temperatures, enabling relatively cheap liquid nitrogen to be used to cool them to the critical temperature. In 1987, J. Georg Bednorz and K. Alex Miller from IBM's Research Laboratory in Zurich received a Nobel prize for their discovery of new ceramic superconductors with exceptionally high transition temperatures.

Material	T_c/K
Zinc	0.87
Mercury	4.15
Lead	7.19
Niobium	9.26
Niobium–germanium	0.23
Yttrium–barium copper oxide	93
Thallium–calcium–bismuth copper oxide	125

table 4.1.3 Some superconducting materials and their critical temperatures.

Questions

1 State Ohm's law.

2 A buzzer in a circuit allows a current of 0.05 A to pass through it when the pd across it is 6 V. What is the buzzer's resistance?

3 What is the difference between an ohmic and a non-ohmic conductor?

4 Graphs of current against voltage are usually drawn with V on the x-axis as the experimental data would be collected by having pd as the independent variable. Why might it be more useful to plot these data with pd on the y-axis?

5 In a resistance experiment, a gold ring was connected into a circuit as a resistor. The connections touched on diametrically opposite points on the circular ring. Its diameter is 2 cm, and the metal's cross-section is a rectangle 3 mm by 0.5 mm. A voltmeter connected across the ring measured 8.3 mV, whilst the current through it was measured at 18 A.

a Calculate the resistivity of gold.

b Describe a practical difficulty with undertaking an experiment of this kind.

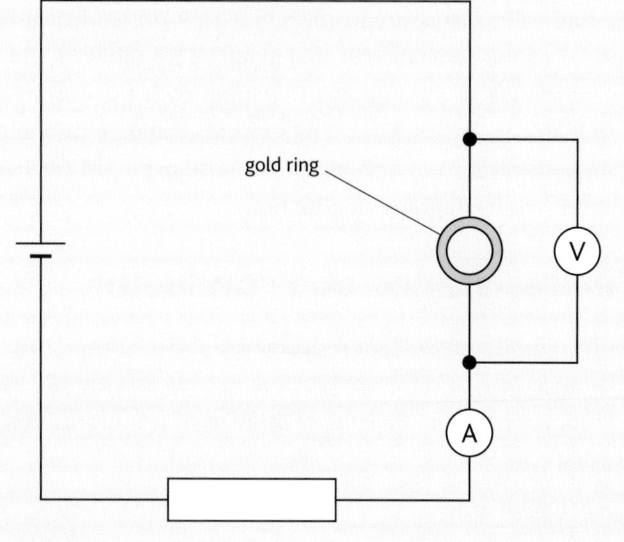

fig. 4.1.14

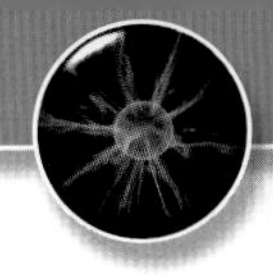

The transport equation

Based on the model of a conductor as a material containing charges which are free to move, an expression can be obtained for the drift velocity of the charges. We can use this to make an estimate of its value.

Consider the piece of conducting material shown in **fig. 4.1.15**.

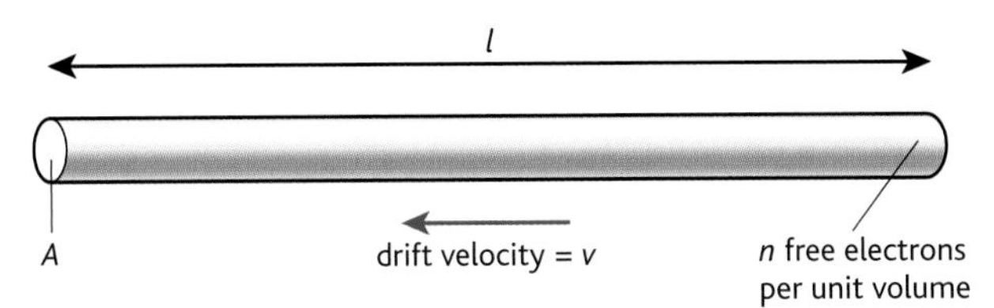

fig. 4.1.15

From the information about the conductor, we may deduce that:

volume of conductor = Al

number of free electrons = nAl

total charge on free electrons = $Q = nAle$

If the current through the conductor is I, we know that $I = Q/t$, so:

current through conductor = $I = \frac{nAle}{t}$

But the drift velocity of the electrons, v, is given by $v = l/t$, so:

current through conductor = $I = nAve$

This is the transport equation.

The transport equation can be written as $I = nAvq$ in order to generalise it for any charge carriers, not just electrons.

Worked examples

Consider a piece of copper and a piece of silicon with exactly the same cross-sectional area of $3 \times 10^{-6}\,m^2$. If both the copper and the silicon carry a current of 100 μA, what is the drift velocity of the electrons in each? Assume that n_{copper} is $10^{29}\,m^{-3}$ and $n_{silicon}$ is $2.6 \times 10^{18}\,m^{-3}$.

$$I = nAve$$

Rearranging:

$$v = \frac{I}{nAe}$$

Substituting the respective values for copper and silicon gives:

$$v_{copper} = \frac{10^{-4}}{10^{29} \times 3 \times 10^{-6} \times 1.6 \times 10^{-19}}$$

$$= 2 \times 10^{-9}\,m\,s^{-1}$$

$$v_{silicon} = \frac{10^{-4}}{2.6 \times 10^{18} \times 3 \times 10^{-6} \times 1.6 \times 10^{-19}}$$

$$= 80\,m\,s^{-1}$$

Questions

1 What current would flow through a copper wire of diameter 0.4 mm if the electrons have a drift velocity of $4.7 \times 10^{-4}\,m\,s^{-1}$?

2 What is the electron density in aluminium wire if it has a diameter of 0.22 mm and carries a current of 3.5 A, with an electron drift velocity of $3 \times 10^{-9}\,m\,s^{-1}$?

4.2 Complete electrical circuits

Power and work in electric circuits

If we consider a lamp in a circuit, the potential difference across it measures the energy transferred to the lamp per unit charge flowing through it, while the current measures the rate of flow of charge through it:

$$V = \frac{W}{Q} = \frac{E}{Q} \qquad I = \frac{Q}{t}$$

If we multiply the pd and the current together we get the rate at which energy is transferred to the lamp:

$$V \times I = \frac{E \times Q}{Q \times t} = \frac{E}{t}$$

or

$$P = VI$$

The rate at which energy is transferred to an element like a lamp in a circuit is called the **power dissipation**. It is measured in **watts**.

Use of the relationship between V, I and R gives us two other relationships for power dissipation:

$$P = VI \text{ and } V = IR \qquad \text{so} \qquad P = (IR) \times I = I^2R$$

and

$$I = \frac{V}{R} \qquad \text{so} \qquad P = V \times \frac{V}{R} = \frac{V^2}{R}$$

Worked examples

a A light bulb is marked '3.5V, 0.1A'. What will be the power dissipated in the bulb when the pd across it is 3.5 V?

$P = VI = 3.5 \times 0.1 = 0.35$ W

b The resistance of a length of wire is measured as 1.5 Ω. At what rate will electrical energy be transferred to heat energy in the wire when a current of 2 A flows through it?

$P = I^2R = (2)^2 \times 1.5 = 6$ W

c A light bulb is marked '12 V, 0.4 A'. How much energy will it transfer if it is on for 20 minutes?

$W = VIt = 12 \times 0.4 \times 20 \times 60$

$= 5760$ J

Work done

Power is the rate at which energy is transferred, so we can find out the total amount of energy transferred – or the work done – by multiplying the equation for power by the time that the device operates for:

$$P = VI$$

$$P \times t = VI \times t$$

$$W = VIt$$

This is an alternative way of combining the equations that define voltage and current:

$$V = \frac{W}{Q} \qquad I = \frac{Q}{t}$$

$$Q = \frac{W}{V} \text{ and } Q = It$$

$$\therefore \frac{W}{V} = It$$

$$W = VIt$$

Questions

1 Explain how Ohm's law can be used to change the equation $P = VI$ into a form which is useful if you do not know the pd involved, but do know the resistance.

2 Calculate the power of the following electrical appliances:

  a a mobile phone which runs on 3.6 V d.c. and uses a current of 0.018 A

  b a mains hairdryer which draws a current of 11 A

  c a torch bulb which uses 6 V and a resistance of 30 W

Circuits containing resistors

Investigating power dissipation

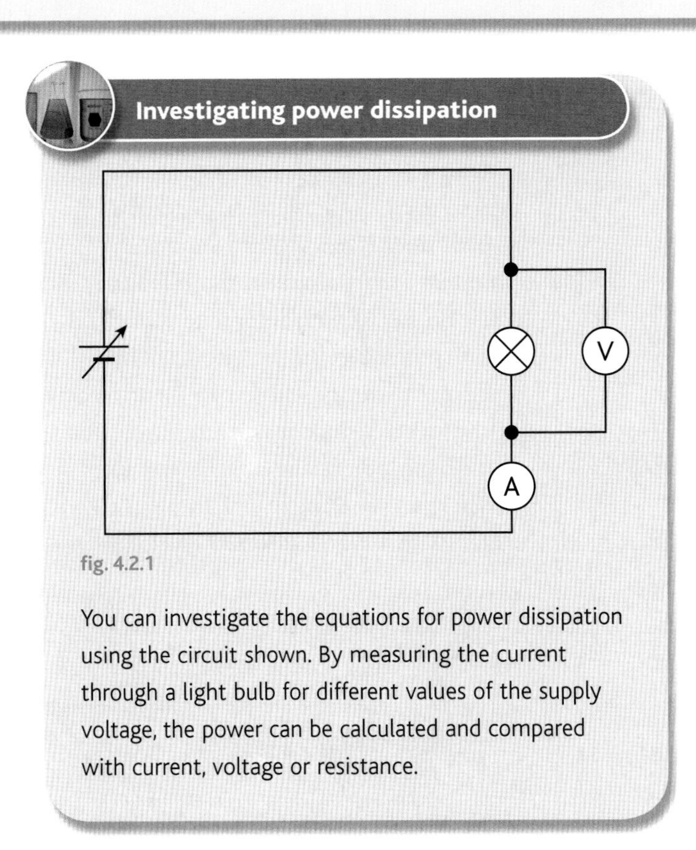

fig. 4.2.1

You can investigate the equations for power dissipation using the circuit shown. By measuring the current through a light bulb for different values of the supply voltage, the power can be calculated and compared with current, voltage or resistance.

Currents in series and parallel circuits

Given three lamps, a source of emf (say three cells) and connecting wire we might want to know how the current flowing round a circuit made of these **circuit elements** behaves.

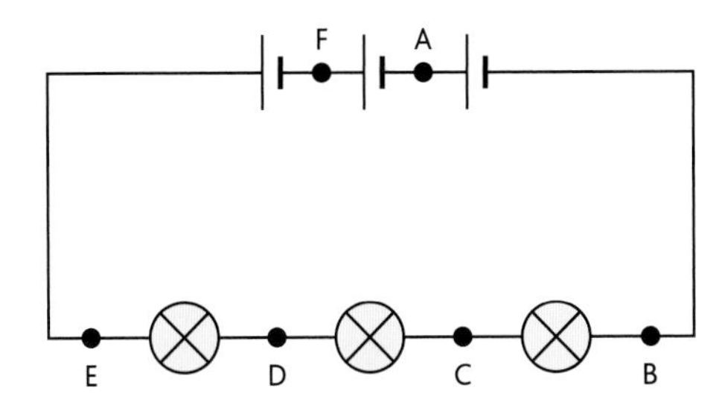

fig. 4.2.2 A series circuit.

Fig. 4.2.2 shows one of the ways in which we can connect these components. The lamps are connected in **series**, so that the current flows first through one and then the other (notice that the cells are connected in series too). What is the size of the current flowing at the different points A to F round the circuit? An ammeter inserted in the circuit at these points gives the same reading at each point. This is what we would expect from the fluid model for electric current discussed in chapter 4.1, since a flow model implies that whatever is flowing is **conserved** (remains constant) as it travels round the circuit.

An alternative way to connect up the lamps is shown in fig. 4.2.3. Here the lamps are said to be in **parallel**.

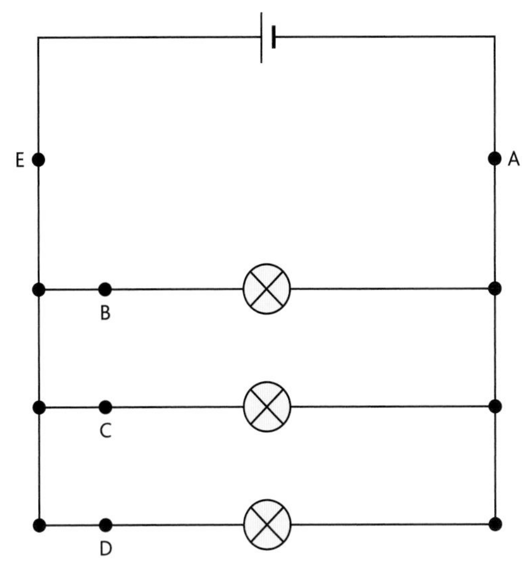

fig. 4.2.3 In this parallel circuit only one cell is used. This causes the lamps to light with the same brightness as when they were connected in series using three cells.

In the circuit in fig. 4.2.3 an ammeter inserted at points A to E in turn will show currents I_A, I_B, I_C, I_D and I_E flowing at these respective points. Analysis of the results of such an investigation shows that:

$$I_A = I_E = I_B + I_C + I_D$$

This is once again in agreement with a flow model for electric current. This behaviour is described by **Kirchhoff's first law**:

The total current into any point in a circuit is equal to the total current out of that point.

Kirchhoff's first law is a result of the **conservation of charge**. All charges flowing into a point must come out again – they cannot mysteriously disappear or appear.

Potential differences in circuits

Having examined the way in which the size of an electric current varies as it flows round a circuit, we shall now examine the variation of potential difference between points round the circuit.

Consider once more the circuit in **fig. 4.2.2**. To measure the pd across each lamp in turn we should need to connect a voltmeter between points B and C, then between points C and D, and finally between points D and E. This would give three readings of pd that we may call V_{BC}, V_{CD} and V_{DE} respectively. Measuring the pd between B and D and between C and E shows that:

$$V_{BD} = V_{BC} + V_{CD}$$

and

$$V_{CE} = V_{CD} + V_{DE}$$

from which it follows that:

$$V_{BE} = V_{BC} + V_{CD} + V_{DE}$$

Experiment confirms this, and shows that the same rule is obeyed when the pd across the cells is measured.

These results show that there is a simple relationship between pds in a circuit, and that they can be regarded as 'adding up' as one goes round a circuit. However, this is not quite the full story, as some careful thought about energy changes in the circuit reveals.

What are the energy changes as current flows round the circuit in **fig. 4.2.4**? The cells transfer electrical energy to the charge flowing through them, and this energy is then transferred in turn to the lamps as the charge flows through them. Clearly more energy cannot be transferred to the lamps than is transferred from the cells, and if no energy is transferred to the connecting wires in the circuit (which will be the case if the resistance of the wires is negligible), then the energy transferred from the cells must equal the energy transferred to the lamps. Since pd measures the energy transferred per coulomb of charge, this must mean that the pd across the lamps is the same as the pd across the cells.

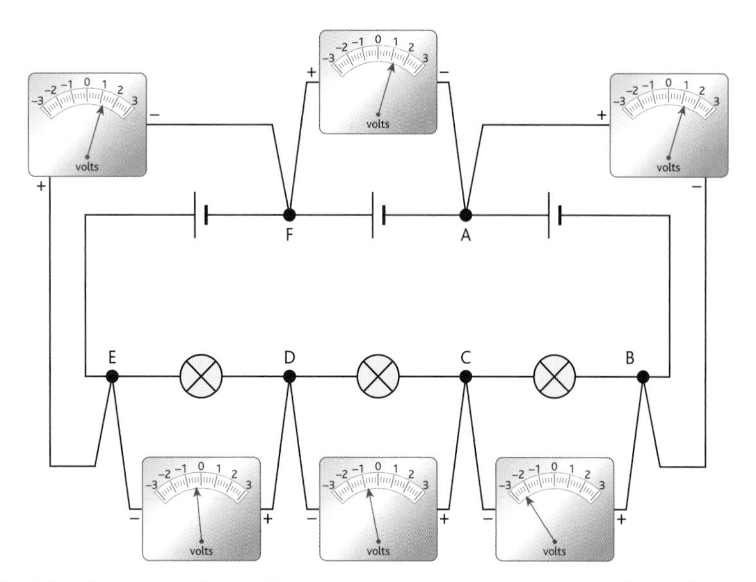

fig. 4.2.4 Going round the circuit with a centre-zero voltmeter produces the results shown here. This shows that, as expected from considering energy flows, the total fall in voltage as one goes from B to E across the lamps is the same as the rise in voltage as one goes from E to B across the cells.

We can draw an analogy between an electrical circuit and a circular walk, as shown in **fig. 4.2.5**. We can regard there as being a certain **potential** at point B in the electrical circuit, and another potential at point C. Between points B and C there is therefore a **potential difference**. If we wish to state the potential at any point in the circuit it must be stated with reference to a fixed point somewhere in the circuit. This fixed point can be chosen to be anywhere in the circuit, since it is used purely for comparing potentials. This is just like the heights on a map, which are always stated with reference to sea level.

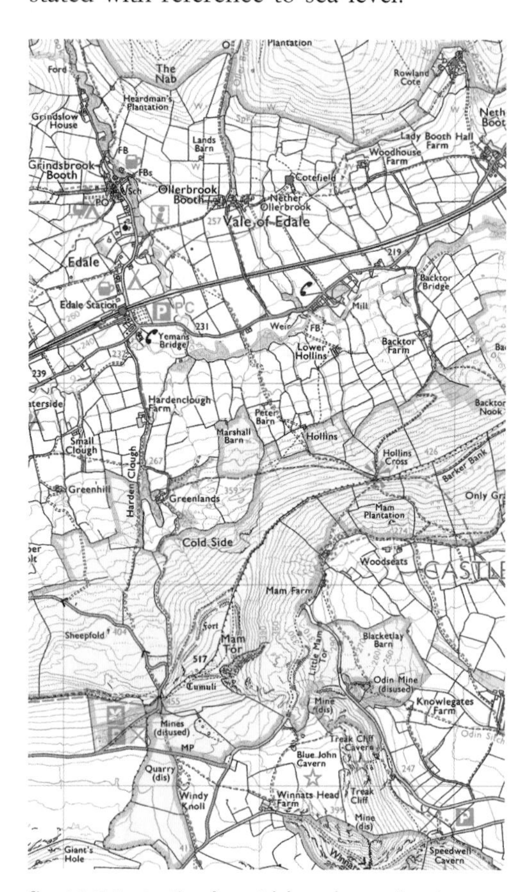

fig. 4.2.5 In starting from Edale and returning there we should be most surprised to end up at a different height above sea level!

As current flows from B to E through the lamps there is a *drop* in potential, as energy is transferred from the flowing charge to the lamps. From E to B there is a *rise* in potential, as energy is transferred from the cells to the flowing charge. The fall in potential is the same

as the rise in potential as we are dealing with a closed circuit – just like the circular walk above. This result is known as **Kirchhoff's second law**:

> **The sum of potential rises and falls around a closed path in a circuit is zero.**

Kirchhoff's second law then is a result of the **conservation of energy**.

Note that the falls in potential occur where energy is transferred *from* the flowing charge, while the rises occur where energy is transferred *to* the flowing charge.

Resistors in series and parallel

When resistors are connected together in a circuit, it is often useful to be able to calculate their combined resistance. This can be done using the principles discussed earlier in this section.

Since the three resistors are connected together in series, we know that they must all have the same current, I, flowing through them. We also know (from Kirchhoff's second law) that the sum of the pds across the individual resistors must be equal to the total pd across all three resistors, that is:

$$V_{total} = V_1 + V_2 + V_3$$

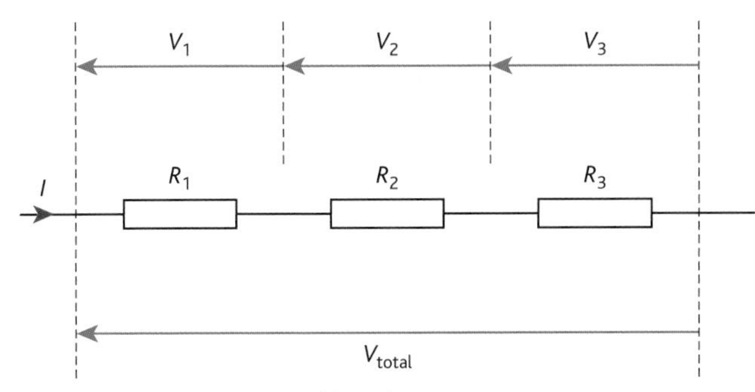

fig. 4.2.6 Resistors connected in series.

Dividing both sides of this equation by I gives:

$$\frac{V_{total}}{I} = \frac{V_1}{I} + \frac{V_2}{I} + \frac{V_3}{I}$$

We know that Ohm's law defines the resistance of each resistor as:

$$R_1 = \frac{V_1}{I} \qquad R_2 = \frac{V_2}{I} \qquad R_3 = \frac{V_3}{I}$$

and the total resistance must be defined as:

$$R_{total} = \frac{V_{total}}{I}$$

Comparing these relationships, we can see that:

$$R_{total} = R_1 + R_2 + R_3$$

This relationship applies to any number of resistors connected together in series.

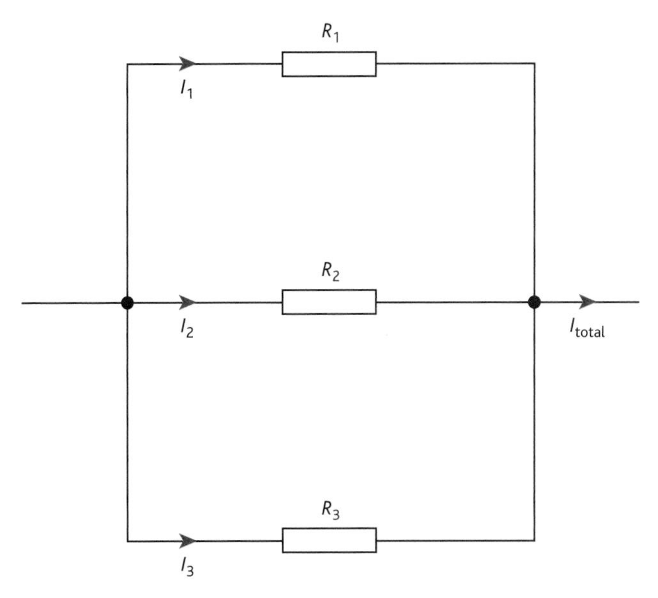

fig. 4.2.7 Resistors connected in parallel.

Because the three resistors are connected together in parallel, they must all have the same potential difference, V, across them. We also know (from Kirchhoff's first law) that the sum of the currents through each individual resistor must be equal to the current through all three resistors in total, that is:

$$I_{total} = I_1 + I_2 + I_3$$

Once again we can apply Ohm's law:

$$I_1 = \frac{V}{R_1} \qquad I_2 = \frac{V}{R_2} \qquad I_3 = \frac{V}{R_3}$$

and the total current can be calculated as:

$$I_{total} = \frac{V}{R_{total}}$$

Comparing these relationships, we can see that:

$$\frac{V}{R_{total}} = \frac{V}{R_1} + \frac{V}{R_2} + \frac{V}{R_3}$$

Dividing both sides of the equation by V gives:

$$\frac{1}{R_{total}} = \frac{1}{R_1} + \frac{1}{R_2} + \frac{1}{R_3}$$

This relationship applies to any number of resistors connected together in parallel.

Worked example

A family with a deaf child has installed a doorbell that also lights a bulb when the bell is rung. The light bulb is connected in parallel with the doorbell's loudspeaker (**fig. 4.2.8**).

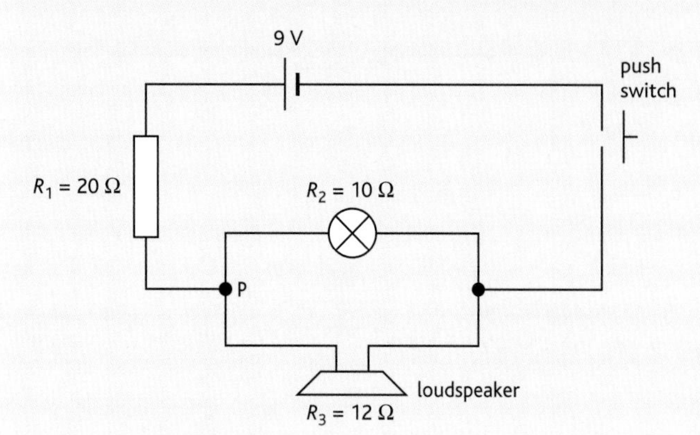

fig. 4.2.8 A doorbell circuit.

How can we calculate the overall resistance in the circuit?

There is a parallel pair of resistances in series with the fixed resistor. The total resistance of the parallel pair is given by:

$$\frac{1}{R_{pair}} = \frac{1}{R_{bulb}} + \frac{1}{R_{speaker}}$$

$$\frac{1}{R_{pair}} = \frac{1}{10} + \frac{1}{12} = 0.100 + 0.083 = 0.183$$

$$R_{pair} = 5.45\,\Omega$$

Now add the series resistor to this:

$$R_{total} = 20 + 5.45 = 25.5\,\Omega \quad \text{(3 significant figures)}$$

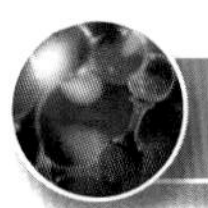

HSW Solving problems in multiloop circuits

Kirchhoff's first and second laws provide us with two powerful tools for analysing the behaviour of circuits in which there is more than one loop and where the solution is not immediately obvious. As an example of this, think about the doorbell circuit from the worked example above. The doorbell loudspeaker will blow if more than 0.2 amps flows through it. With the addition of the bulb, what will be the current through the speaker now?

Consider the various currents and potential differences around the circuit as shown in **fig. 4.2.9**.

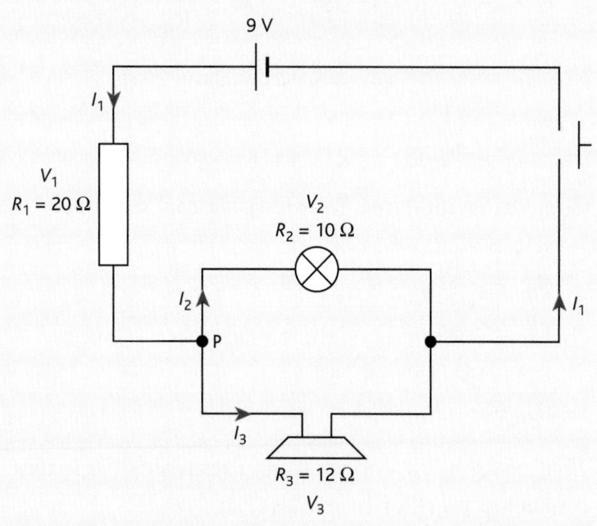

fig. 4.2.9 A circuit with two loops

As a first step we assign currents to the branches, which are shown as I_1, I_2 and I_3 in **fig. 4.2.9**. In addition, we name the potential difference over each resistance – V_1, V_2 and V_3.

Applying Kirchhoff's first law at the junction P:

$$I_1 = I_2 + I_3 \qquad \text{equation 1}$$

Remember that $V_2 = V_3$, so that Kirchhoff's second law gives us:

$$V_1 + V_3 = 9\,\text{V}$$

$$I_1R_1 + V_2 = 9\,\text{V} \qquad \text{equation 2}$$

From above we have that total resistance in the circuit $R_{total} = 25.5\,\Omega$, so:

$$I_1 = \frac{V_{total}}{R_{total}} = \frac{9}{25.5} = 0.353\,\text{A}$$

Substituting I_1 and R_1 into equation 2:

$$(0.353)(20) + V_2 = 9V$$

$$V_2 = 9 - 7.06 = 1.94V \ (= V_3)$$

We can now use Ohm's law on the loudspeaker on its own:

$$I_3 = \frac{V_3}{R_3}$$

$$I_3 = \frac{1.94}{12} = 0.16\,\text{A}$$

So the current through the loudspeaker is less than 0.2 amperes and it will operate properly.

Questions

1 Draw a circuit diagram to show how a simple battery-powered torch might be wired up. Describe and explain how the current measurements would compare if ammeters were simultaneously placed in the middle of each of the wires shown on your diagram.

2 Explain why Kirchhoff's second law is an example of the conservation of energy.

3 You have three 50 Ω resistors. These can be wired in series and parallel combinations to make four different total resistance values. Draw the possible different wiring combinations and calculate the total resistance for each combination.

The potential divider

The principle of the **potential divider** is an important one in many applications that use circuits containing resistors. The volume control knob on a music system is a potential divider. Circuits that sense changes using devices like light-dependent resistors and thermistors use potential dividers too.

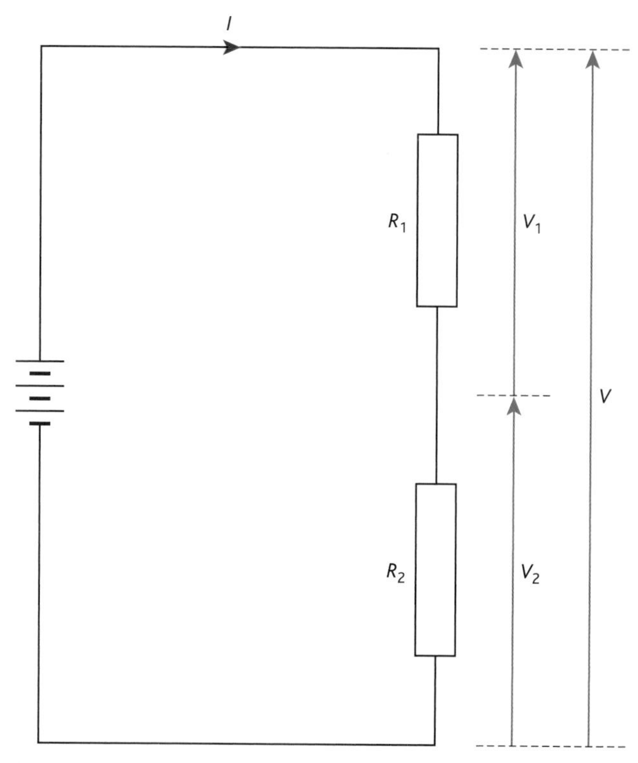

fig. 4.2.10 The voltages on this diagram rise from bottom to top, as shown by the single-headed arrows.

A potential divider is a device that consists essentially of two resistors. A current I flows through the two resistors, and the effect of this is to split or divide the potential difference across the resistors in two. A simple example of a potential divider is shown in **fig. 4.2.10**. Using Ohm's law we may write

$$V_1 = IR_1 \text{ and } V_2 = IR_2$$

Dividing the first equation by the second gives us

$$\frac{V_1}{V_2} = \frac{R_1}{R_2}$$

which means that the total potential difference, V, across the two resistors has been divided up in the ratio of their two resistances – hence the name potential divider. By choosing appropriate values of R_1 and R_2 any voltage between zero and V can be obtained across either of the two resistors.

Worked example

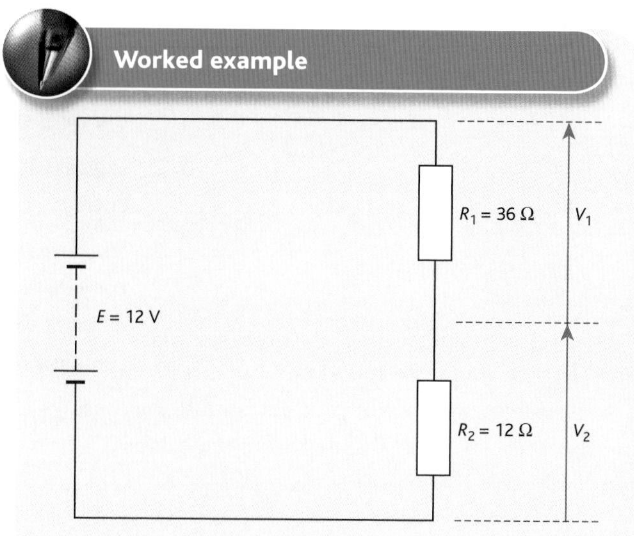

fig. 4.2.11

Fig. 4.2.11 shows a battery with an emf of 12 V with a pair of resistors R_1 and R_2 connected to it. R_1 has a resistance of 36 Ω, while R_2 has a resistance of 12 Ω. What is the pd across each resistor?

The total pd across the two resistors is 12 V. This is divided up in the ratio of their two resistances:

$$\frac{V_1}{V_2} = \frac{R_1}{R_2} = \frac{36}{12} = 3$$

Since:

$$V_1 = 3V_2 \text{ and } V_1 + V_2 = V$$

We know that:

$$3V_2 + V_2 = V \text{ so } V_2 = \frac{V}{4}$$

In this case this gives:

$$V_2 = \frac{12}{4} = 3\text{V}$$

and $V_1 = 3V_2 = 3 \times 3 = 9\text{V}$

The potentiometer

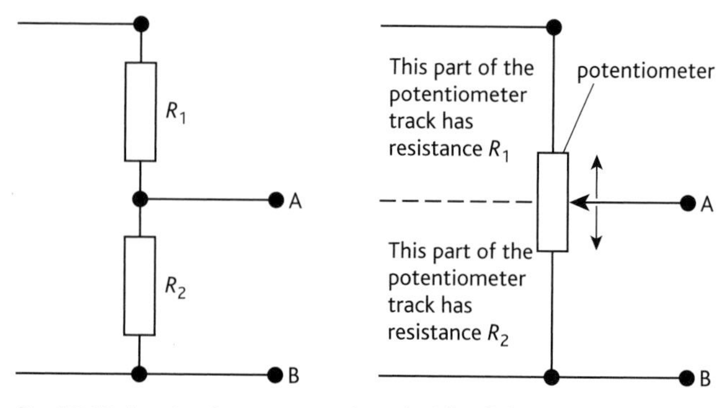

fig. 4.2.12 A potentiometer uses the principle of the potential divider circuit.

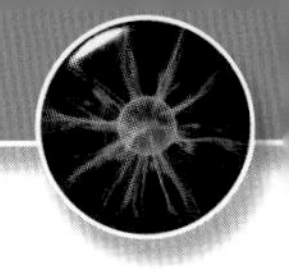

Where a continuously variable pd is required, the fixed resistors in the potential divider circuit are replaced by a **potentiometer**. This is a device that consists of a length of resistance wire or a carbon track, along which a wiper may be moved – effectively a resistor with a sliding contact. Fig. 4.2.12 shows that as the wiper is moved along, the values of R_1 and R_2 alter, although $R_1 + R_2$ remains constant. In this way, the pd appearing across terminals AB can be continuously varied from zero up to the emf of the supply. Practical potentiometers come in different shapes and sizes, depending on the job they have to do (fig. 4.2.13). An example might be the volume control for a music system, which will produce sound loudness in proportion to the pd supplied to the speakers.

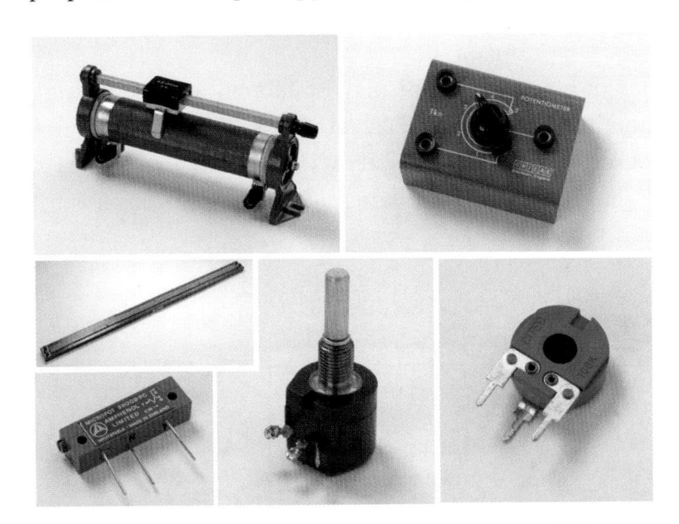

fig. 4.2.13 The photos show some potentiometers, but at widely differing scales. The slide wire is a 1 m length of resistance wire. The other photos show miniature potentiometers used in electronic circuits.

Questions

1 a A series circuit is constructed using a 3000 Ω resistor, a 2000 Ω resistor and a 12 V dc supply. If a voltmeter were connected across the 3000 Ω resistor, what pd would it show?

b If the 2000 Ω resistor is replaced by one that is 6000 Ω, what would the new voltmeter reading be?

2 A variable resistor is made from a metre of resistance wire. The resistance of the whole length of wire is 80 ohms. If you had a 6 V battery and wanted to provide 3.7 V to power your iPod directly, how would you connect up this variable resistor and battery?

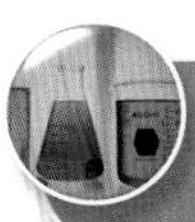

Investigating variation of resistance with length of conductor

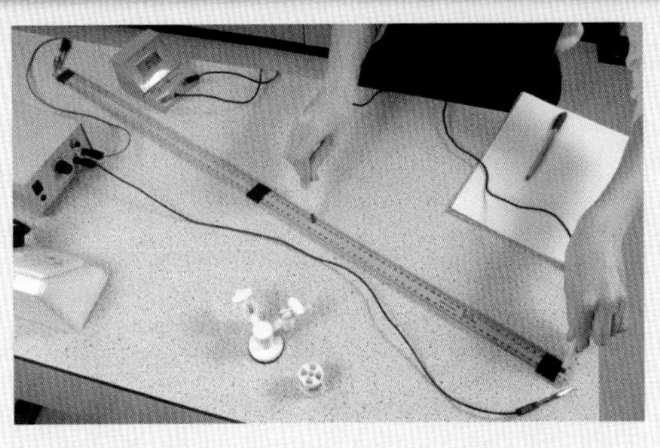

fig. 4.2.14 Set-up for variation of resistance investigation

You can investigate how the resistance of a wire varies with its length by measuring the pd across different lengths of a fixed piece of wire. This will show you the basic setup of a potentiometer.

You could use this setup to investigate the resistivity equation further, if you were to plot the resistance against length. These quantities are related by the formula:

$$R = \frac{\rho l}{A}$$

So a straight line produced from these results would have a gradient equal to ρ/A, which would allow you to calculate the resistivity of the metal involved.

We have seen that the factor influencing the pd developed across the terminals of a potential divider or a potentiometer is the *ratio* of the two resistances, not the *absolute* value of either resistance. What determines the actual values that might be used in practical circuits?

The answer to this question depends on the current to be drawn from the potential divider. The current drawn from the device should not be an appreciable fraction of the current through the resistors, otherwise the assumption that the current through R_1 and R_2 is the same will not be valid. This means that if an appreciable current is required, a large current must flow through the potential divider itself. This may well be undesirable, since this current does nothing useful, and the energy it carries will largely be transferred to the resistors of the potential divider, making them hot. This is seldom a problem, however. The usual applications for potential dividers often involve electronic sensing circuits and do not draw large currents, and in practice there is little difficulty in using large resistors that lead to a small current through the divider itself, while providing any voltage required between zero and the emf of the supply.

Sources of emf – internal resistance

Sources of emf – internal resistance

A source of emf always has some resistance to electric current within it, called its internal resistance. The **internal resistance** of a source of emf has two effects:

1. it results in the voltage across the terminals (**terminal potential difference**) of the source dropping as a current is drawn from it
2. it results in the source being less than 100% efficient as energy is dissipated in the internal resistance as current flows through it.

The voltage quoted on the label of a source of emf like a battery is the voltage measured when no current is being drawn from it – often called the **open circuit voltage**.

The internal resistance of a source of emf may be thought of as a resistor r in series with the nominal emf E, as **fig. 4.2.15** shows. When a current I is drawn from the source a pd ΔV appears across the internal resistance r, so that $\Delta V = Ir$. The voltage V across the terminals of the source falls, so that:

$$V = E - \Delta V = E - Ir$$

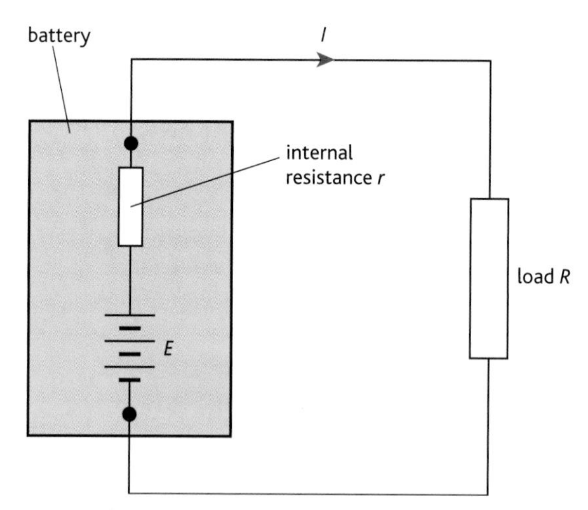

fig. 4.2.15 Circuit diagram showing a battery with internal resistance.

fig. 4.2.16 The lights dim when the key is turned to start the engine.

The internal resistance of a source of emf results in energy being dissipated within the source itself. This effect can be seen quite clearly when starting a car on a cold, dark morning – turning the starter motor draws a current that may be as large as several hundred amps, producing a significant voltage drop across the internal resistance of the battery. If the car is started with the headlights on, they can clearly be seen to dim as the pd across them drops.

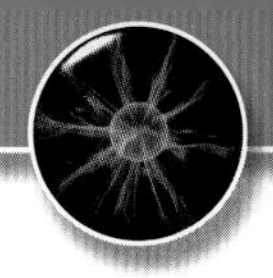

The internal resistance of a source of emf sets limits on the amount of power that can be supplied to an external circuit or load. A source of emf connected to a load will dissipate some of its energy in its internal resistance, and transfer the rest to the load. We often want to transfer as much electrical power as possible from a source to a load, for example when an amplifier is connected to a loudspeaker, or when an aerial is connected to a television. Investigations show that in order to do this, the resistance of the load must be equal to the internal resistance of the source, since the power supplied by a source to a load varies as shown in fig. 4.2.17.

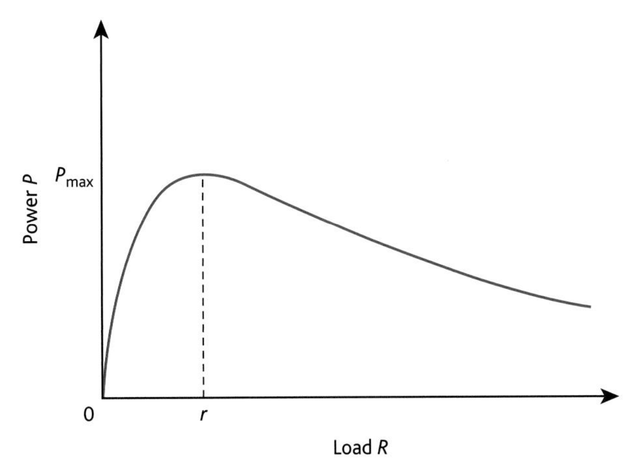

fig. 4.2.17 Graph of power against load. There is a peak in the power transferred when the load R is equal to the internal resistance r.

Questions

1 Why is the emf of a power supply sometimes different from the reading on a voltmeter connected across its terminals?

2 In an experiment to investigate internal resistance, the circuit below was used to find the current and voltage for several settings of the variable resistor. How could you draw a graph of the results to be used to find the emf and internal resistance of the cell?

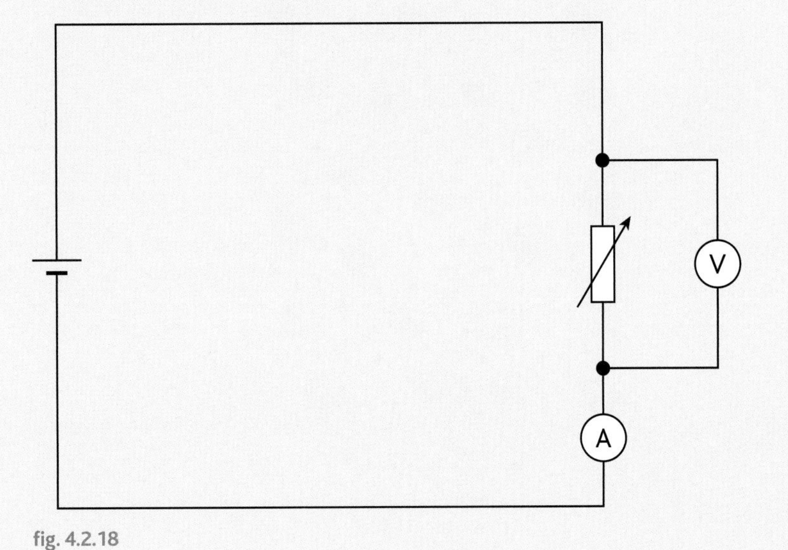

fig. 4.2.18

Understanding conduction

We began this topic by introducing an electric current as a flow of charge. Having used a flow model to derive the relationships we have seen so far, it is now time to consider the mechanisms by which charge might flow through conductors, and why charge is unable to flow through insulators.

Conduction in metals

One possible explanation for the high thermal conductivity of metals is the idea that the outermost electrons in the atoms of a metal are not bound to their atoms and form a sort of 'sea', in which these electrons are free to move around through the body of the metal. They can carry heat energy and so such a model can explain thermal conductivity quite convincingly, and may also be used to explain the electrical conductivity of metals.

The free electrons in a metal have a random motion like the motion of the particles of a gas – in fact the electrons in a metal are often referred to as making up an **electron gas**. At room temperature the electrons have a velocity of around $10^5 \mathrm{m\,s^{-1}}$.

Any solid is made up of an arrangement of atoms – usually called a lattice – bonded together. When a source of emf is connected to the ends of a conductor this causes an electric field through the conductor (**fig. 4.2.19**). This field has the effect of causing the electrons in the conductor to move with an average velocity known as the **drift velocity**. The size of this drift velocity depends on a number of factors, but is typically around $10^{-7} \mathrm{m\,s^{-1}}$.

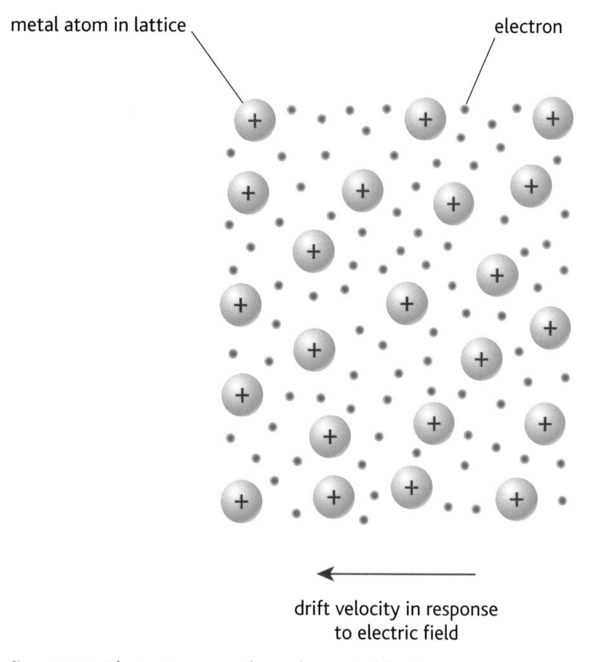

fig. 4.2.19 Electrons move through a metal lattice in response to an electric field caused by a source of emf.

Why is this value so much smaller than the actual velocity of the electrons given above? Although the individual electrons each have an instantaneous velocity of many thousands of metres per second, they collide with the atoms fixed in the lattice and bounce back, so their overall movement is slow – like moving three steps forward and two steps back (**fig. 4.2.20**). Collisions with lattice atoms is the basic idea used to explain resistance. This also explains why the resistance of a metal wire increases as it gets hotter (**fig. 4.2.21**). The lattice atoms will vibrate faster and further when hotter, and this increases the number of collisions with the conduction electrons, reducing the current flow.

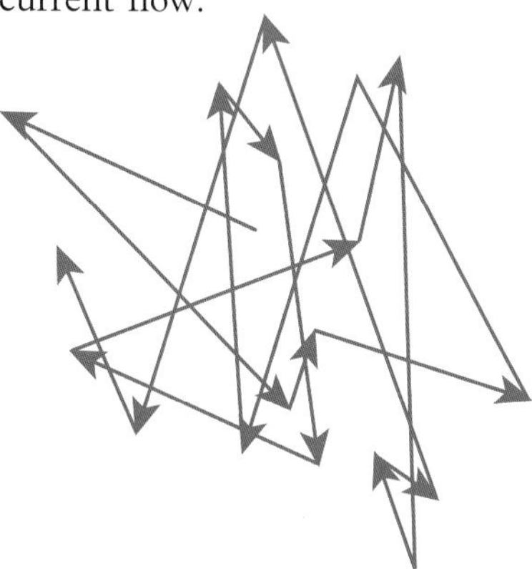

fig. 4.2.20 Collisions reduce the velocity of electrons in a metal lattice.

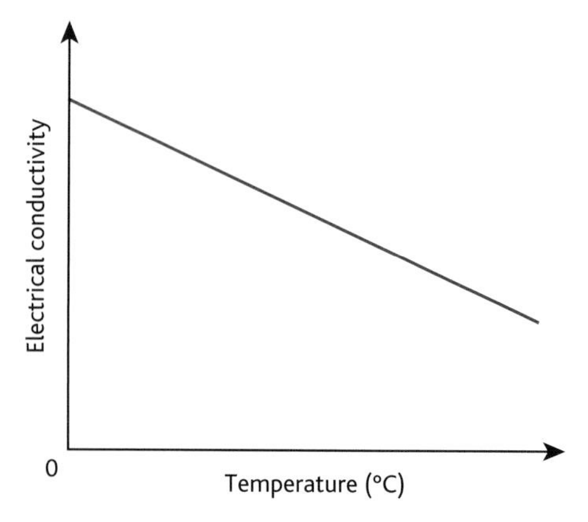

fig. 4.2.21 The electrical conductivity of copper falls as its temperature rises. This behaviour is common to most metals.

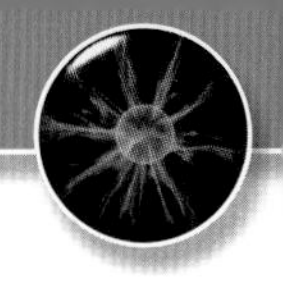

Conduction in pure semiconductors

Substances like silicon and germanium have resistivities between those of insulators and those of conductors, as shown in **table 4.1.2** (see page 125). These substances are known as **semiconductors**, and they form the basis of many of the devices that we take for granted in a technological society. Pure semiconductors are usually referred to as **intrinsic** semiconductors, since their conductivity is not affected by any external factors. As we shall see below, trace impurities can greatly alter the conductivity of semiconductors.

In an intrinsic semiconductor electric current is carried by moving electrons, as in metals. In the section on the transport equation in chapter 4.1, we saw that the number of charge carriers in silicon is perhaps a billion times fewer than in copper. However, in addition to electrons, intrinsic semiconductors can also be considered to contain moving *positive* charges that carry current. This can be explained as follows.

Some of the electrons in an atom of an intrinsic semiconductor are held less tightly than others. This means that in a piece of intrinsic semiconductor at room temperature there will always be a few free electrons that have been 'shaken free' of their atoms by thermal excitation – they have absorbed energy from their surroundings. When an electron leaves an atom in this way, the atom becomes positively charged. The effect of an electron leaving an atom is therefore to create a positive charge in the semiconductor lattice. This positive charge is called a **hole**. When an electric field is applied to the semiconductor (that is, when it is connected to a source of emf) the electrons and holes move in opposite directions, and the semiconductor exhibits **intrinsic conduction** because the charge carriers have arisen inside the conductor.

The movement of electrons and holes in the lattice is shown in **fig. 4.2.22**. Under the influence of an electric field, electrons move through the lattice. Electrons still bound to the atoms in the lattice are able to migrate from an atom to a nearby hole, thus causing the hole to appear to move through the lattice. This motion happens in the opposite direction to the motion of the electrons.

An analogy for the movement in holes through the lattice is shown in **fig. 4.2.23**.

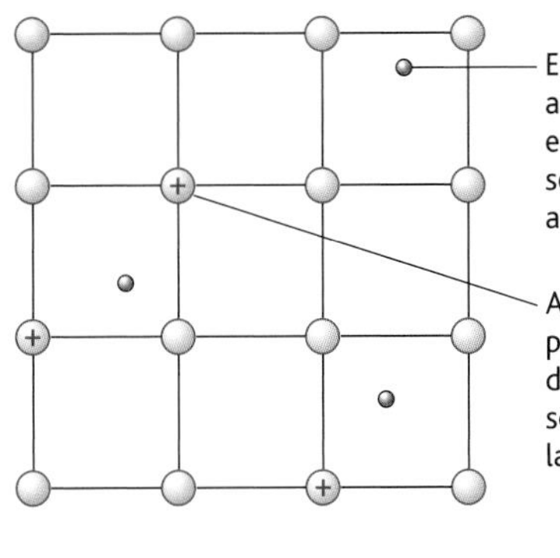

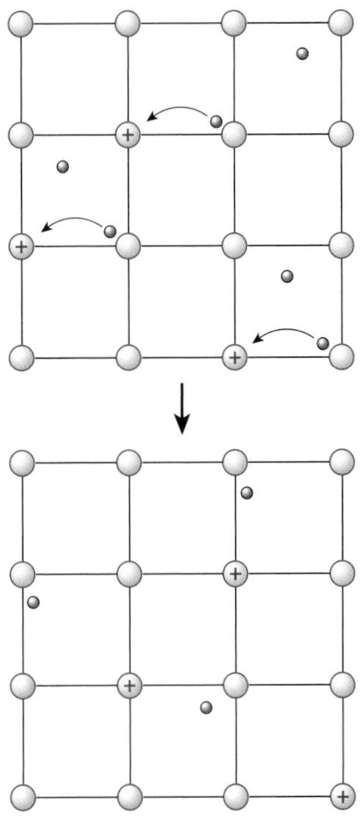

fig. 4.2.22 The current in a pure semiconductor consists of free electrons moving through the semiconductor lattice in one direction, with an equal number of positively charged holes moving in the other direction.

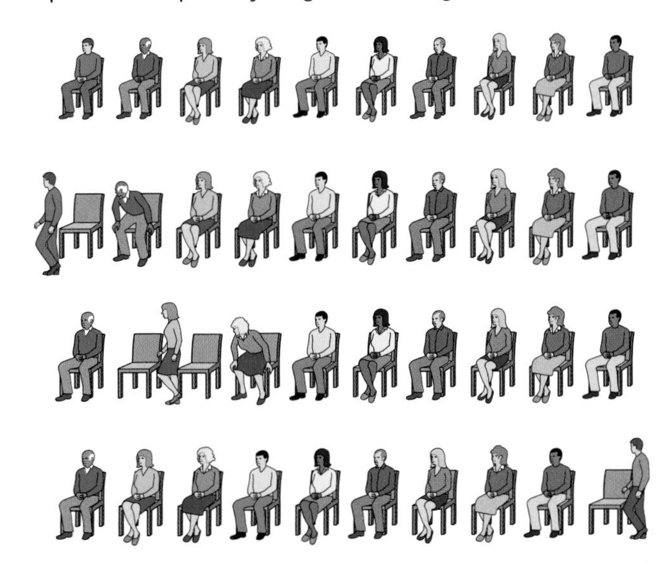

fig. 4.2.23 The motion of holes through the lattice is like the motion of the empty chair in this dentist's waiting room. As the people move from right to left, the empty chair moves from left to right.

If the temperature is raised, more electrons in the semiconductor will be 'freed' and the conductivity increases (the resisitivity goes down). This behaviour leads to the use of semiconductors in an interesting device for measuring temperature. The thermistor (an abbreviation of 'thermal resistor') is an electronic component that indicates temperature by the value of its resistance. NTC thermistors have a negative temperature coefficient. This means that as the temperature goes up, their resistance goes down. As the temperature goes up, more conduction electrons are released, so the current increases, meaning that the resistance has effectively reduced (see fig. 4.1.10, page 125).

Calibrating a thermistor

Calibration of the thermistor must be done very carefully, as each semiconductor chemical has different properties in terms of the heat energy needed to free conduction electrons. You can calibrate a thermistor yourself by measuring its resistance in a beaker of water as you vary the water temperature.

fig. 4.2.24 Calibrating a thermistor.

Conduction in 'doped' semiconductors

The use of semiconductors in devices like transistors and diodes and in integrated circuits depends on altering their conducting properties by introducing minute quantities of impurities in a process called doping. Doping results in an **extrinsic** semiconductor, because the impurity introduces extra charge carriers to the semiconductor lattice. Doping introduces extra charge carriers by replacing atoms in the semiconductor lattice with atoms of an impurity of similar size (this is important so as to minimise the distortion of the lattice that occurs).

The number of free electrons and holes can be altered dramatically by the addition of tiny quantities of impurity. For example, the addition of only one arsenic atom per million silicon atoms increases the conductivity 100 000 times. Arsenic is an example of a **donor** impurity, which releases free electrons into the lattice, increasing the number of negative charge carriers and so producing an n-type semiconductor. Boron on the other hand is an example of an **acceptor** impurity, which traps electrons when introduced into the lattice, resulting in an increase in the number of positive holes. A semiconductor of this type is called a p-type semiconductor. Fig. 4.2.25 shows how these processes occur.

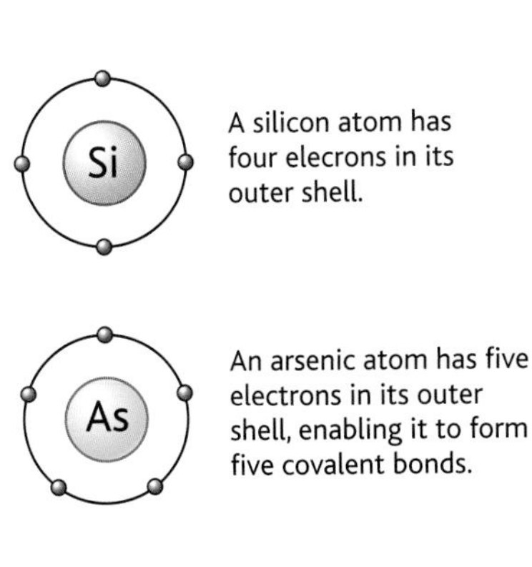

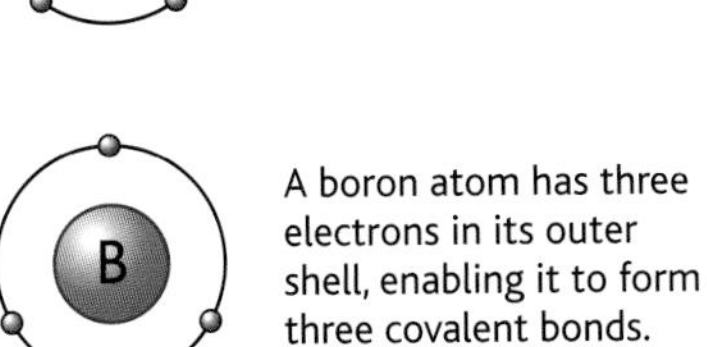

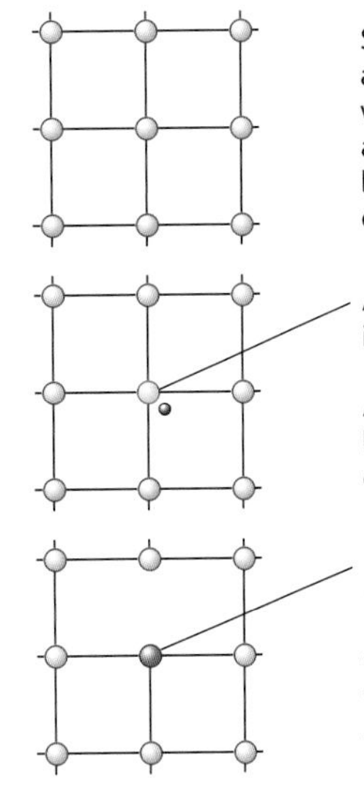

fig. 4.2.25 Semiconductor doping.

Conduction in insulators

An **insulator** can be thought of as an extreme version of a semiconductor. In a semiconductor, heat energy excites electrons from atoms in the lattice and enables them to escape from the atom and become free to carry charge. The energy required to do this is sufficiently small that this process is possible at room temperature. In an insulator, the energy required to free an electron from an atom is much greater than for a semiconductor. As a result, an insulator has so few free electrons that its conductivity is very small indeed.

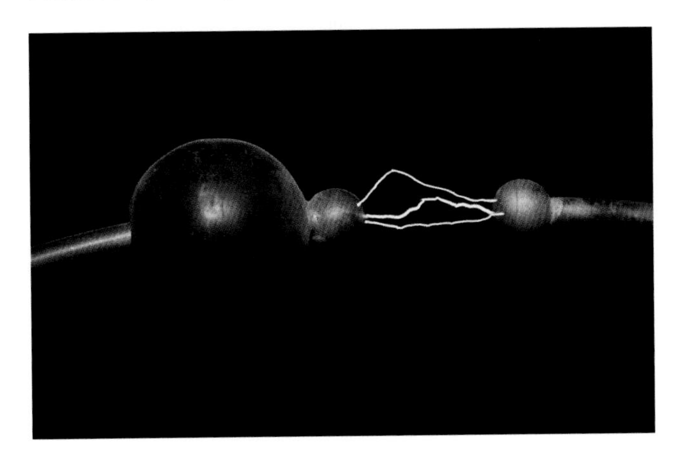

fig. 4.2.26 Insulators can conduct!

There are two circumstances in which an insulator may be made to conduct electricity. Fig. 4.2.26 shows two switch contacts on the National Grid being closed. Under these circumstances a small air gap between the contacts has a very large pd across it. The few free charge carriers that exist in this air are accelerated by the large pd and acquire enough kinetic energy to ionise any atoms they hit. These in turn are accelerated and ionise other atoms, resulting in an **avalanche** of charged particles that act as charge carriers and cause the air to become conducting. When a glass rod is heated to red heat, a pd across the rod is then able to cause a current to flow, as shown by the lit bulb in the circuit. In this situation the high temperature of the glass in the rod is sufficient to cause some of the atoms in the glass to lose electrons that then become available for conduction. *Do not try this experiment yourself* – it requires high temperatures and large pds.

Questions

1. a Draw a graph to show how the temperature of a light bulb filament affects its resistance.
   b Explain the shape of your graph in terms of conduction electrons and lattice vibrations.
   c How does your answer in part **b** relate to the transport equation?
   d How does your graph illustrate the possibility of a superconductor?
2. Explain what a 'hole' is in electrical conduction in silicon.
3. What is the basic difference between an n-type and a p-type semiconductor doping element?
4. How does doping increase the conductivity of some semiconductors?
5. Explain why the resistance of some semiconductors goes down when they get hotter, despite the fact that their lattice vibrations increase with temperature.

Sensing and control circuits

HSW Automation

Many developments in our modern society have machines functioning by themselves. Street lamps come on when it is dark, but only when it is dark enough. Central heating turns on when it is too cold and, critically, turns off again when the house is warm enough. These self-controlling devices must have the ability to sense the environment and make a decision as to what to do based on the sensations. Resistors that vary depending on the conditions, such as a thermistor that has a lower resistance when it is hotter, can be combined with a potential divider circuit in order to perform these automated functions.

A potential divider circuit could be used inside a refrigerator to switch on the cooling circuit when the temperature becomes too high (more than 3 °C).

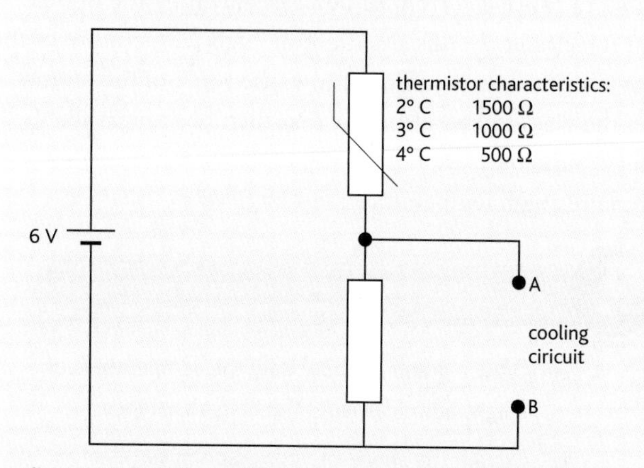

fig. 4.2.27 Possible temperature control circuit for a refrigerator.

If the cooling circuit needs a pd of 5 V or more between A and B in order to operate, we can check that it will come on if the temperature goes above 3 °C.

$$V_{AB} = \frac{6 \times 5000}{5000 + R_t}$$

At 2 °C:

$$V_{AB} = \frac{6 \times 5000}{5000 + 1500} = 4.6\text{ V}$$

so the cooling circuit is off.

At 3 °C:

$$V_{AB} = \frac{6 \times 5000}{5000 + 1000} = 5.0\text{ V}$$

so the cooling circuit is on.

At 4 °C:

$$V_{AB} = \frac{6 \times 5000}{500 + 500} = 5.5\text{V}$$

so the cooling circuit is on.

Problems with load circuits

In connecting a load (some resistance) up to a potential divider, care must be taken that the load does not upset the calculations done before connecting the load up to the divider. In the refrigerator example above, suppose the cooling circuit has a resistance of 5000 Ω. Then the parallel combination of the fixed resistor and cooling circuit would have a combined total resistance of 2500 Ω. Now the calculations become:

$$V_{AB} = \frac{6 \times 2500}{2500 + R_t}$$

At 2° C:

$$V_{AB} = \frac{6 \times 2500}{2500 + 1500} = 3.8\text{ V}$$

so the cooling circuit is off.

At 3° C:

$$V_{AB} = \frac{6 \times 2500}{2500 + 1000} = 4.3\text{ V}$$

so the cooling circuit is still off.

At 4° C:

$$V_{AB} = \frac{6 \times 2500}{2500 + 500} = 5.0\text{ V}$$

so the cooling circuit is on.

Thus, the cooling circuit now does not come on until the temperature reaches 4 °C, which may be too warm for storing some perishables.

Selecting a voltmeter

Voltmeters must be connected in parallel across the two points in a circuit between which we wish to know the potential difference. The connecting of this measuring device should not significantly alter the current flowing through the circuit in any way. In order for this to be so,

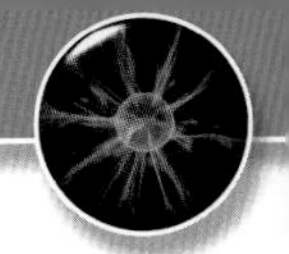

the voltmeter should have as large a resistance as possible in order that as little current as possible should flow through it.

Consider some measurements to be made on a potential divider using two voltmeters, a situation shown in **fig. 4.2.28**.

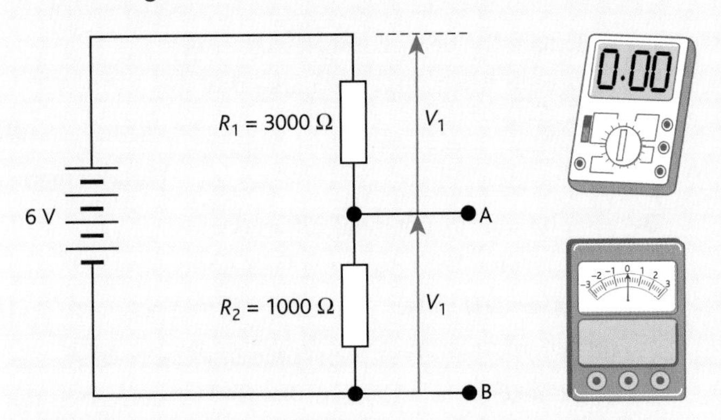

fig. 4.2.28

To begin with, the meters are connected to the terminals of the battery in turn – both give a reading of 6.0 V. When the electronic multimeter is connected to terminals AB, it reads 1.5 V as expected. However, when the other meter is connected to the terminals, it reads only 1.0 V.

The key to understanding this apparent paradox is to realise that connecting the voltmeter between two points affects the potential difference between them. Although the battery has a resistance itself (see the section on internal resistance of power supplies earlier in this chapter), this is low compared with the resistance of both meters. However, the resistance of the laboratory meter R_v is comparable to that of R_2 in the potential divider, and so it alters V_2 significantly. This is explained in **fig. 4.2.29**.

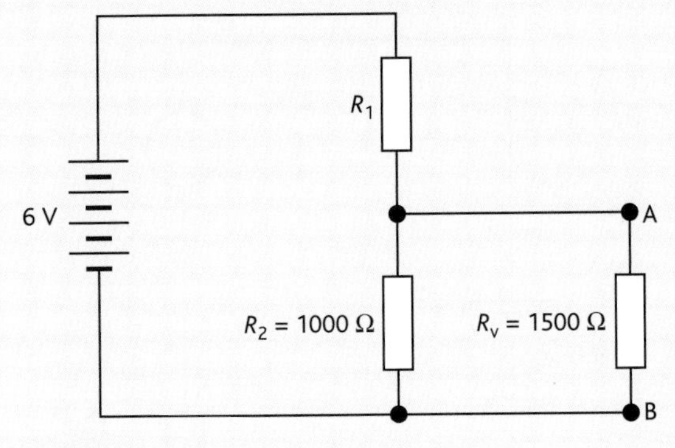

fig. 4.2.29

With the meter connected across terminals AB, the resistance between these two points drops to R_{new}:

$$\frac{1}{R_{new}} = \frac{1}{1000} + \frac{1}{1500} = \frac{5}{3000}\ \Omega^{-1}$$

so

$$R_{new} = 600\,\Omega$$

This means that:

$$\frac{V_1}{V_2} = \frac{R_1}{R_2} = \frac{3000}{600} = 5$$

Therefore V_2 is now 1.0 V.

This does not happen with the electronic voltmeter, since it has a resistance of around $10^{10}\,\Omega$. Connecting this meter across AB changes V_2 by only about one part in ten million, which is virtually undetectable! To ensure that a voltmeter does not appreciably alter the pd between two points in a circuit it should have a resistance around 10^3 times greater than the resistance of the circuit element(s) between the two points.

The properties of systems

As a general rule, it is impossible to make an observation of some property of a system without disturbing it in some way. The examples using voltmeters that we have just seen show that this is so for electrical circuits – and it is true for other systems too.

If we wish to measure the pressure of the air in a car tyre, for example, it is necessary to connect a pressure gauge to the tyre in order to draw off some of the air to enable the gauge to measure the pressure in the tyre. Similarly, a thermometer must be brought into thermal equilibrium with the system that is to be measured, which results in a transfer of energy from system to thermometer or vice versa. In all these cases the disturbance to the system can be minimised so that the disturbance is negligible in comparison to other uncertainties involved in the measurement, but this does not negate the fact that the disturbance is there.

In all but the most sensitive experiments in the macroscopic world such problems are easily overcome. On a microscopic scale however, things are much more difficult. How to observe an electron without disturbing it may seem a trivial problem, but it is one of many similar ones that has exercised great minds during the course of the last century, even giving rise to a creature called Schrödinger's cat! You can research on the internet to see if Schrödinger's cat is dead yet.

Examzone: Topic 4 DC electricity

1 A negative temperature coefficient thermistor is used in the following circuit to make a temperature sensor.

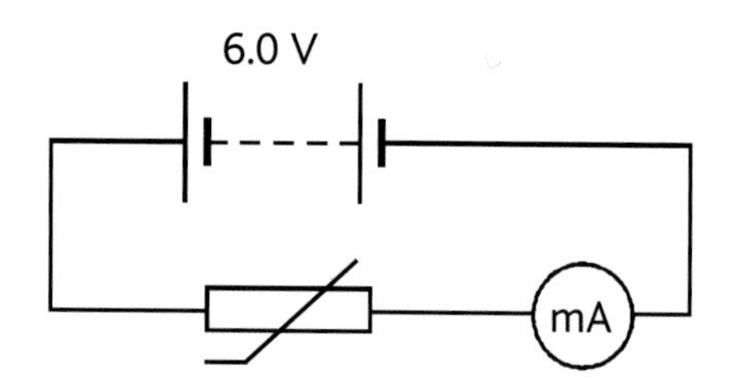

Explain how the circuit works. **(2)**

The graph shows how the resistance of the thermistor varies with temperature.

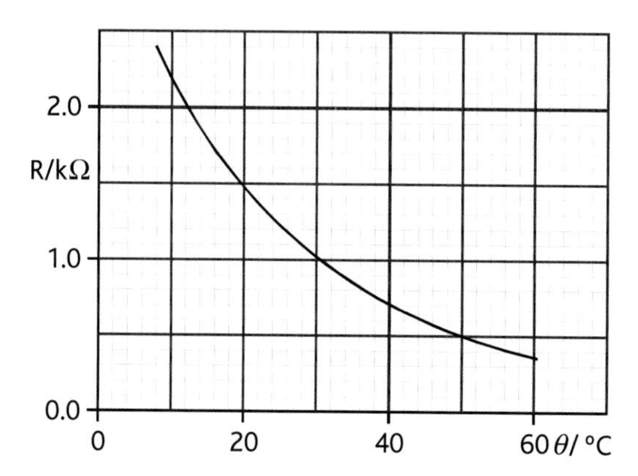

Look at the graph and write down what the reading on the milliammeter will be when the thermistor is at a temperature of 20 °C? **(3)**

(Total 5 marks)

2 a i) Write the word equation that is used to define charge. **(1)**

ii) Write down what is potential difference. **(1)**

b A 9.0 V battery of negligible internal resistance is connected to a light bulb.

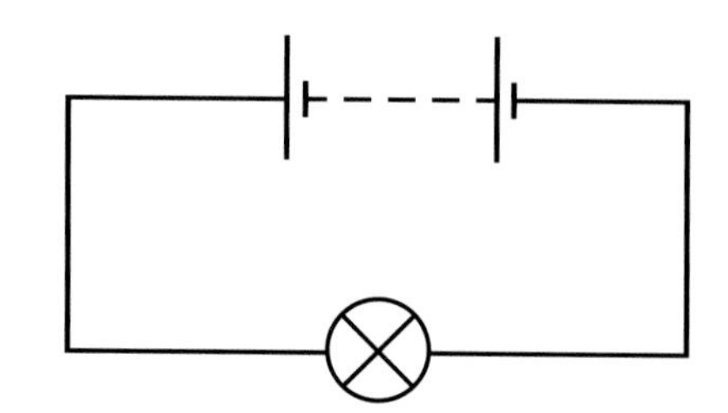

Calculate the energy transferred in the light bulb when 20 C of charge flows through it. **(2)**

(Total 4 marks)

3 a The unit of potential difference is the volt. Express the volt in terms of base units only. **(1)**

b A 6.0 V battery of negligible internal resistance is connected to a filament lamp. The current in the lamp is 2.0 A.

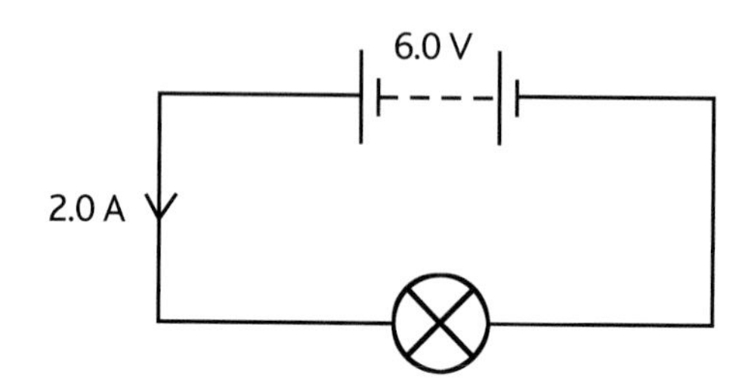

Calculate how much energy is transferred in the filament when the battery is connected for 2.0 minutes. **(3)**

(Total 4 marks)

4 a A student sets up a circuit and accidentally uses two voltmeters V_1 and V_2 instead of an ammeter and a voltmeter. The circuit is shown below.

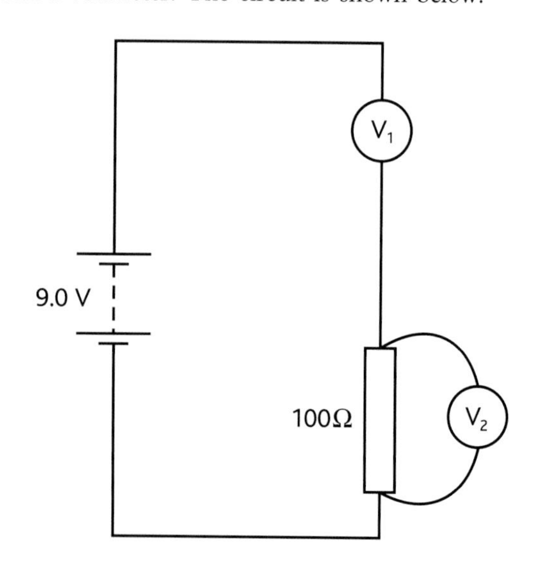

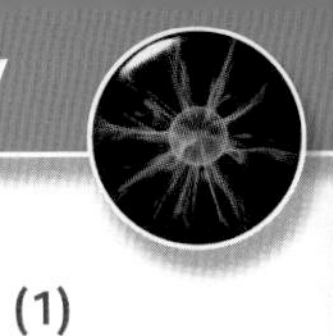

(i) Write down which voltmeter should be an ammeter. **(1)**

(ii) Both voltmeters have a resistance of 10 MΩ. The student sees that the reading on V_2 is 0 V. Explain why the potential difference across the 100 Ω resistor is effectively zero. **(2)**

b The student replaces the 100 Ω resistor with another resistor of resistance R. The reading on V_2 then becomes 3.0 V.

(i) Copy and complete the circuit diagram below to show the equivalent resistor network following this change.

Label the resistor R. **(2)**

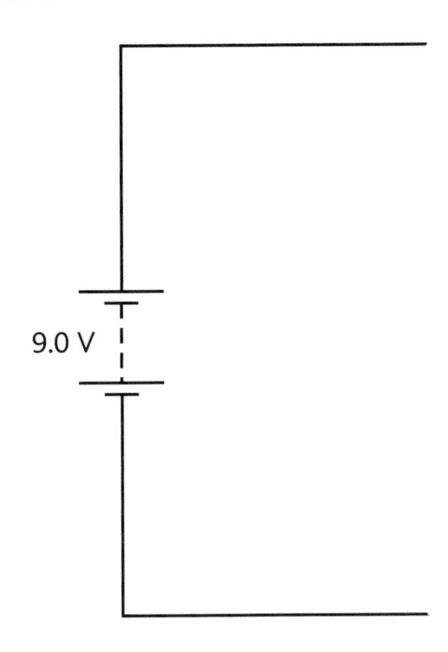

(ii) Calculate the value of R. **(3)**

(Total 8 marks)

Topic 5 Nature of light

Observing a sunbeam on a sunny day may make light seem like a simple thing. Not only is it subtle and complex, but determining its exact nature is still eluding scientists. In this topic you will learn of some of the characteristics of light which are straightforward, but you will also be introduced to some of the more wondrous aspects. In particular, you will see how light can sometimes be considered to be a wave but, under other conditions, it can act as a particle.

What are the theories?

Light is a wave. Light is a particle. Light is a wavicle? Unit 2 Topic 3 looked at many situations in which light acts as a wave. Here we will spend more time looking at the situations in which it acts as a particle, called a photon. This will allow us to consider how light is produced, and how much energy it carries. In particular, the photoelectric effect highlights the particle nature of light.

What is the evidence?

When scientists discovered the photoelectric effect in the early twentieth century, it shattered the wave theory which had so well explained all previous observations of light. Einstein produced an equation to describe this effect, and you will able to practise using this equation, even if you do not have access to experimental equipment which demonstrates photoelectricity.

What are the implications?

Scientists have long debated the nature of light, often through investigations of vision. The historical pendulum of belief has swung backwards and forwards from wave to particle and back again several times throughout history, and you will be introduced to some of the more important proponents of each way of thinking.

Understanding the nature of light, albeit to the limited extent that scientists have managed, allows us to use it as a source of energy in solar cells. These are becoming ever more efficient, and ever cheaper, as our understanding of light improves. As the sunshine landing on Earth every day provides an unimaginably large amount of energy, harnessing this cheaply and efficiently with solar cells is currently one of the most important areas of research.

The map opposite shows you all the knowledge and skills you need to have by the end of this topic. The colour in each box shows which chapter they are covered in and the numbers refer to the sections in the Edexcel specification.

Chapter 5.1

recognise and use the expression $E = hf$ (part of 67)

explain how the behaviour of light can be described in terms of waves and photons (63)

recall that the absorption of a photon can result in the emission of a photoelectron (64)

understand and use the terms threshold frequency and work function and recognise and use the expression $hf = \phi + \frac{1}{2}mv^2_{max}$ (65)

define and use radiation flux as power per unit area (69)

explain how wave and photon models have contributed to the understanding of the nature of light (71)

use the non-SI unit, the electronvolt (eV) to express small energies (66)

explore how science is used by society to make decisions – remote sensing (part of 72)

Chapter 5.2

recognise and use the expression $E = hf$ to calculate the highest frequency of radiation that could be emitted in a transition across a known energy band gap or between known energy levels (67)

explain atomic line spectra in terms of transitions between discrete energy levels (68)

explore how science is used by society to make decisions – the viability of solar cells as a replacement for other energy sources (part of 72)

recognise and use the expression efficiency = [useful energy (or power) output]/[total energy (or power) input] (70)

5.1 What is light?

A brief history of light

The debate about the nature of light has raged from the very earliest of times. The Ancient Greek philosopher Democritus believed that objects were visible because the atoms of which they were made 'swarmed' into the air and entered the observer's eyes – in other words, a particle theory of light. Other philosophers of the time disagreed. Empedocles argued that objects became visible when touched by light rays emitted from the eyes, while Plato held that the eyes emitted light rays that intercepted the light rays emitted by objects. This debate, polarised into almost opposing views about the nature of light and vision, has continued right up to the present century.

The first person to propose a wave model for light was Leonardo da Vinci, in the late fifteenth century, comparing the reflection of light to the echoing of sounds. This theory was given support by the work of the Italian physicist Francesco Maria Grimaldi. A paper published in 1665 after his death described his experiments with the diffraction of light, and set out his ideas about light as a wave motion. In the same year, Robert Hooke compared light to waves in water, and Isaac Newton discovered that a prism may be used to split white light into the colours of the rainbow.

The end of the seventeenth century saw a fierce debate about the nature of light. Newton argued that light was a stream of particles, and showed how the properties of reflection and refraction could be explained using a particle model. He also argued that if light behaved as a wave it should form 'fuzzy' shadows around objects rather than the sharp-edged shadows that one sees on a sunny day. (Newton was probably unaware of the earlier work of Grimaldi, and certainly did not realise that the wavelength of light is too small to show such behaviour with any but the smallest everyday objects.) In opposition to this view, the Dutch physicist Christiaan Huygens described in detail the wave theory of light in his *Traité de la lumière* (Treatise on light), including the principle of secondary wavelets used in chapter 3.2 of this book.

By around 1700 there was good evidence to suggest that light had properties that could best be explained in terms of a wave motion. This view was supported by eminent scientists of the day like Hooke and Boyle, who used it to explain the formation of colours in oil floating on the surface of water in terms of the interference of two waves. Newton rejected these ideas however, and for the next century or so the particle theory of light was generally accepted by scientists.

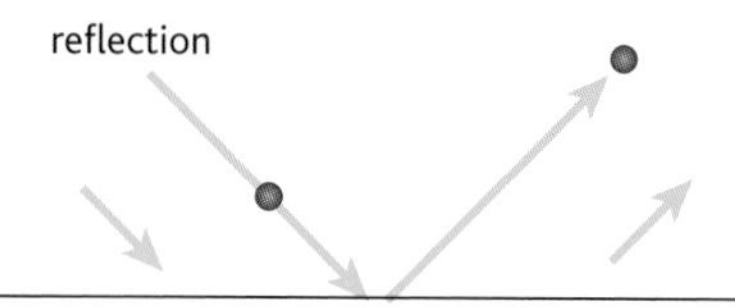

'Particle of light bounces off surface with component of velocity reversed and horizontal component unchanged (compare this with a ball bouncing off a hard surface at an angle).

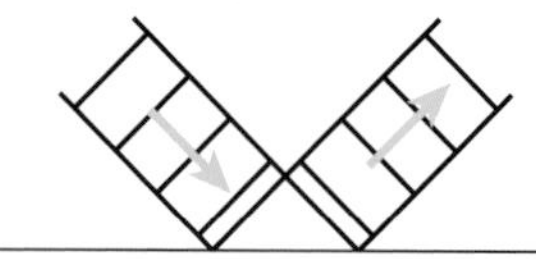

Light wave is reflected as surface acts as a source of secondary wavelets (see chapter 3.2).

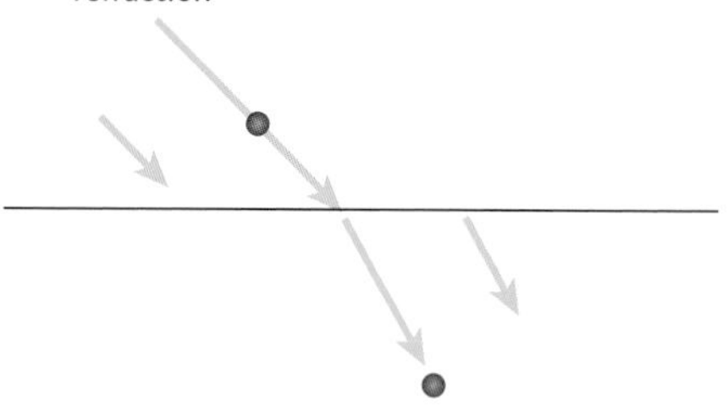

'Particle' of light enters optically denser medium. Horizontal component of velocity is unchanged, but vertical component is increased.

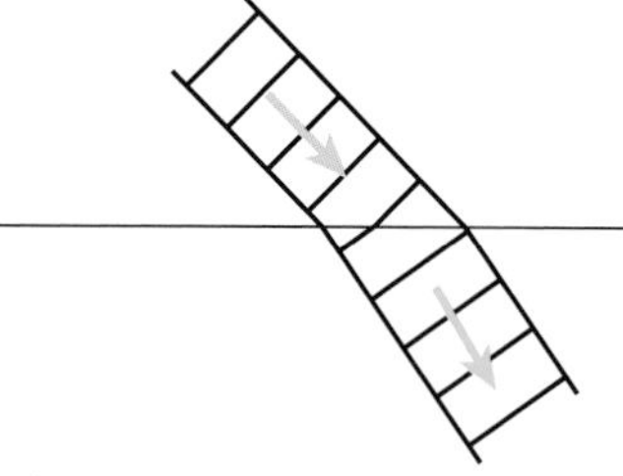

Light wave enters optically denser medium and slows down (velocity in all directions is affected).

fig. 5.1.1 **Reflection and refraction explained using Newton's and Huygens' ideas. There is no easy way of testing the two theories to compare them in the case of reflection. Refraction however can be used to test them – provided that the speed of light in an optically dense medium like glass or water can be measured.**

In 1802, Thomas Young published *On the theory of light and colours*, in which he set out his ideas about light, giving his account of two-slit interference (see chapter 3.2). The evidence from Young's observations – many of which Grimaldi had recorded nearly 200 years previously – made it difficult, if

not impossible, to continue to think of light as having a particle nature. This idea was extraordinarily difficult to overthrow however, possibly because Newton himself had argued for it so forcefully. The middle of the century brought proof to convince everyone. In 1853, the French physicist Léon Foucault showed that light travelled more slowly in water than in air. This spelled the death of the particle theory of light, since the theory required that light must travel more quickly in water than in air in order to explain refraction (see **fig. 5.1.1** above).

But Newton was not entirely wrong. In 1905, Albert Einstein showed that the photoelectric effect (see pages 150–151) could only be understood if light were thought of as a stream of particles rather than as a wave. Nearly 20 years later, Louis de Broglie produced his theory of **wave–particle duality**, in which he showed that anything that had particle properties could also be shown to have wave properties too – even objects like tennis balls and people! De Broglie's theory received confirmation in 1927 when two physicists working quite independently (Davisson and Thomson) showed that electrons could be diffracted by crystals. The modern science of **quantum mechanics** regards light as both a stream of particles and as a wave. Sometimes the wave-like properties are more important than the particle-like properties, and sometimes it is the other way around. The particles of light are called **photons**.

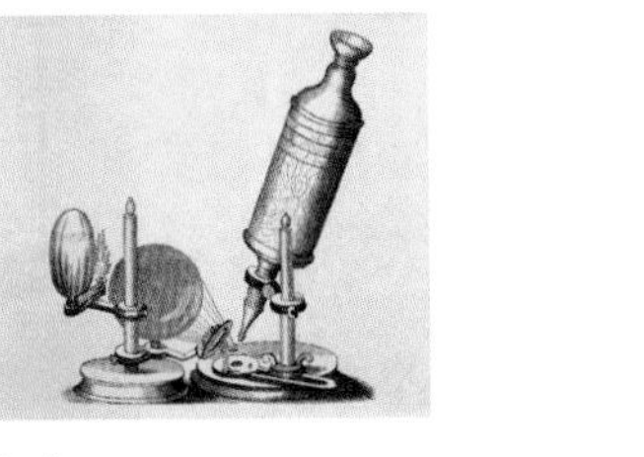

Robert Hooke
1665. In his Micrographia, Robert Hooke set out many of his observations made through a microscope. As well as describing cells in living tissue for the first time, the work contained the idea that light behaves like a wave on water.

Christiaan Huygens
1690. Christiaan Huygens' work set out his wave theory of light.

Isaac Newton
1704. Isaac Newton's work Optics set out his particle theory of light.

Thomas Young
1802. Thomas Young published his work describing the interference of light having passed through two narrow slits. Physicists were still loathe to abandon the particle model of light.

James Clark Maxwell
1864. James Clark Maxwell used Faraday's ideas about fields to explain the nature of light waves as electromagnetic waves. He predicted that other waves with different wavelengths but with the same properties of light were possible.

Albert Einstein
1905. Albert Einstein showed that the photoelectric effect could be understood by treating light as a stream of particles.

Louis de Broglie
1923. Louis de Broglie's theory of wave–particle duality was proposed, and became a central part of the science of quantum mechanics. The idea that light has both particle and wave properties is still accepted at the start of the twenty-first century.

fig. 5.1.2 Some important names and dates in the development of theories about light.

Questions

1 Name two scientists who supported the particle theory of light and two who supported the wave theory.

2 Why was Foucault's experiment to measure the speed of light in water so important?

3 In the eighteenth century, virtually all experimental evidence available pointed to light being a wave, contrary to Newton's views. Most scientists continued to believe light was a particle. Why?

4 Which theory, waves or particles of light, is the one that scientists currently believe is correct?

Wave or particle?

The nature of light

At around the same time as Thomson was investigating the behaviour of cathode rays and coming to the conclusion that they were particles, a German physicist called Max Planck was trying to obtain a theoretical model to understand the way in which a **black body** emits electromagnetic radiation. (A black body is a perfect emitter and absorber of electromagnetic radiation, capable of absorbing and radiating all wavelengths.) He found that this was impossible unless he made an assumption that ran completely contrary to the laws of physics known at the time – he had to assume that energy could be absorbed or radiated by a body only in discrete quantities, not in continuous amounts.

Based on thermodynamics, Planck showed that the emission of thermal radiation by a black body could be modelled using a reasonably simple relationship that included a new constant h, now known as **Planck's constant**. Planck's constant has a value of 6.63×10^{-34} Js. This model pictured the black body as a series of oscillators, rather like 'marbles on a spring'. Planck envisaged that the oscillators vibrated in such a way that they could only absorb or lose specific amounts of energy, although the radiation they absorbed or emitted remained a continuous distribution of energies as demanded by classical physics.

fig. 5.1.3 Max Planck, discoverer of the quantisation of energy. Planck invented the idea of quantisation in 'an act of desperation' because 'a theoretical explanation had to be found at any cost, whatever the price.' He received the Nobel prize for this work in 1918.

Although Planck's work was not fully understood at the time, it was taken up by Einstein. In 1905, Einstein showed that the radiation from a black body could be understood more simply if it was assumed that the radiation itself was **quantised**, consisting of particle-like packets of energy. Each packet is called a photon.

HSW Black body radiation

A black body is a perfect absorber of radiation, able to absorb completely radiation of any wavelength that falls on it. Fig. 5.1.4a shows the simplest possible type of black body – a hole in a box that is painted black inside.

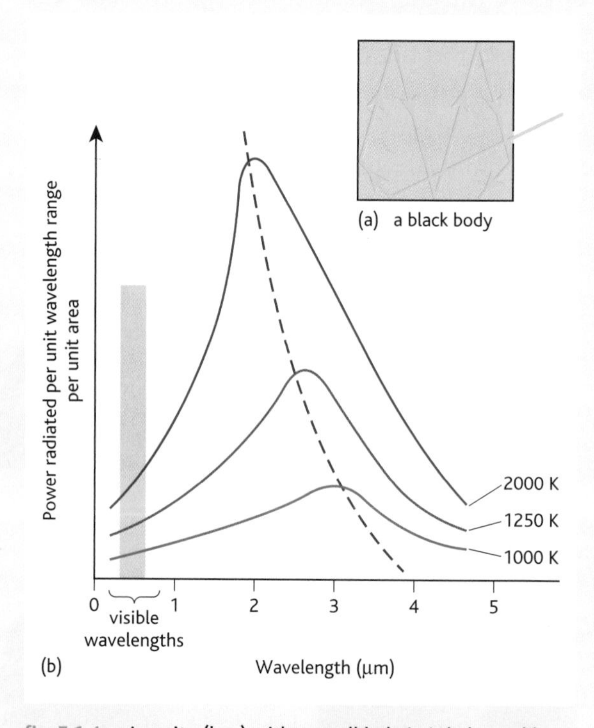

fig. 5.1.4 a A cavity (box) with a small hole in it behaves like a black body. b As the temperature of the black body increases, the peak in power radiation occurs at shorter wavelengths.

A black body not only absorbs all wavelengths of radiation falling on it, but also radiates all wavelengths. The exact spectrum of the radiation from the black body depends on its temperature (fig. 5.1.4b). Classical physics can predict this pattern of emission at long wavelengths but not at short wavelengths, leading to a disagreement that physicists christened the ultraviolet catastrophe.

The peak wavelength of the radiation emitted is inversely proportional to the temperature of the black body (fig. 5.1.4b), a relationship known as Wien's displacement law. An understanding of this relationship is important to astronomers, since stars behave as black body radiators. Knowledge of the wavelengths of the radiation from a star enables astronomers to make estimates of its temperature.

If the frequency of a wave is f, the energy of the photons in it is given by the relationship:

$$E = hf$$

where h is Planck's constant.

Radiation flux

If we know the energy of each photon in a beam, and the number of photons in the beam, we can calculate the rate that it delivers energy. How intensely light shines on something is measured by the quantity **radiation flux**, which is a measure of the amount of energy landing on a unit area in a unit time. Energy in a unit time is the definition of power, which means that radiation flux F is defined as the power P per unit area A:

$$F = \frac{P}{A}$$

Thus, the unit for radiation flux is W m^{-2}.

The fact that light behaves like a wave under some circumstances and like a particle under others demonstrates that neither the wave model nor the particle model of classical physics is adequate for understanding the behaviour of light. For a full understanding of light we have to regard it as a **wavicle**, a wave–particle object that behaves sometimes like a classical wave and at other times like a classical particle – and at other times like a mixture of the two! This is clear if we examine another puzzle from physics at the turn of the century, the **photoelectric effect**.

Worked examples

Energy from the Sun strikes the Earth's surface with a flux of about $1000\ \text{W m}^{-2}$. What rate of arrival of photons is this, assuming that the average wavelength of sunlight is 500 nm?

When basking in the Sun, how much heat energy would this deliver to your face each second?

Take Planck's constant as 6.63×10^{-34} J s and the velocity of light as $3 \times 10^8\ \text{m s}^{-1}$. We know that:

$$c = f\lambda$$

so we can write the energy E of one photon as:

$$E = \frac{hc}{\lambda}$$

Substituting values:

$$E = \frac{6.63 \times 10^{-34} \times 3 \times 10^8}{5 \times 10^{-7}}$$

$$= 4 \times 10^{-19}\ \text{J}$$

The number of photons n arriving per metre^2 per second is given by the energy incident per metre^2 per second divided by the photon energy. Recalling that $1\ \text{W} = 1\ \text{J s}^{-1}$, we have:

$$n = \frac{1000}{4 \times 10^{-19}}$$

$$= 2.5 \times 10^{21} \text{ photons per m}^2 \text{ per second}$$

The 'particle nature' of light is not likely to be evident in our macroscopic world, owing to the vast numbers of photons involved.

Take your face to be a circle of diameter 15 cm, i.e. $r = 0.075$ m. Then:

$$\text{area } A = \pi r^2 = \pi \times (0.075)^2 = 0.0177\ \text{m}^2$$

$$P = F \times A = 1000 \times 0.0177 = 17.7\ \text{W}$$

So your face receives 17.7 J of heat energy per second.

Questions

1 Calculate the energies of the following photons:

  a UV with a frequency of 4×10^{15} Hz

  b X-ray with a wavelength of 8.4×10^{-11} m

  c light at the peak response of the human eye $\lambda = 550$ nm.

2 For the answer to question **1 c**, calculate the number of photons per second that would need to be emitted from a light bulb if it gives off 60 W of light.

3 An iPod battery in use delivers 4.4 W. If the battery was to be replaced by a square handheld solar cell, what would the side length of the solar cell need to be? Assume all sunlight energy landing on the cell is converted into electricity.

4 Briefly describe the 'ultraviolet catastrophe', and what idea Planck had to come up with to reconcile it.

The photoelectric effect

Metals can lose electrons if they are given enough energy. This is shown by the production of electrons by thermionic emission in an electron gun, such as that used in a cathode ray oscilloscope. The electron gun uses heat to supply energy to the electrons, but it is possible to use light too, as the demonstration illustrated in **fig. 5.1.5** shows.

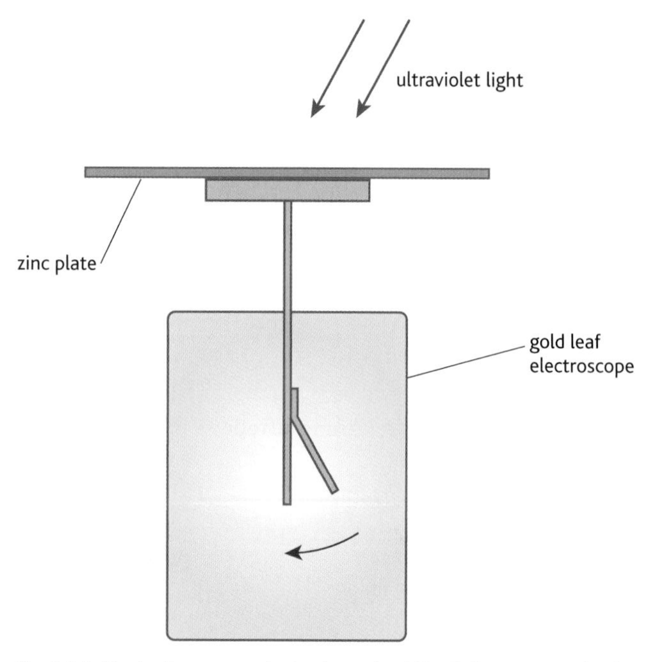

fig. 5.1.5 The leaf on a negatively charged gold leaf electroscope slowly falls if a zinc plate resting on the cap of the electroscope is irradiated with ultraviolet light.

An investigation like this quickly shows a number of interesting features.

1 With no light falling on the zinc plate, the leaf falls only very slowly if at all (due to charge leaking away through the air).

2 Ultraviolet light shone onto the plate causes the leaf to fall rapidly. This fall is stopped if a sheet of glass is placed between the zinc plate and the ultraviolet lamp (glass absorbs ultraviolet light strongly).

3 The rate of fall of the leaf depends on the distance of the lamp from the plate. The closer the lamp, the more rapid the fall.

4 Visible light has no effect on the behaviour of the leaf.

5 A positively charged electroscope is unaffected by ultraviolet light.

Prior to Einstein's work, scientists had interpreted results like this as showing that ultraviolet light was capable of transferring energy to electrons in a metal, giving them sufficient energy to escape from the metal surface. Electrons liberated in this way were given the name **photoelectrons**. The leaf of the negatively charged electroscope falls as electrons are ejected from the zinc plate, decreasing the overall negative charge on the plate. The leaf does not fall if the electroscope is positively charged, as the ultraviolet light cannot transfer electrons to the plate – only energy.

For a given surface there is a minimum frequency of light (the **threshold frequency**) below which no emission of photoelectrons occurs. This accounts for visible light having no effect on the electroscope in **fig. 5.1.5** and why a sheet of glass stops the leaf of the electroscope falling. The number of photoelectrons emitted from the metal surface depends only on the intensity of the light falling on it. Moving the ultraviolet lamp closer to the electroscope therefore increases the rate at which photoelectrons are ejected from the plate, as the intensity of the incident radiation increases.

Even though physicists could agree on the general interpretation of the photoelectric effect, using classical physics no model could be constructed that could explain:

a why there was a threshold frequency below which no photoelectrons were emitted, and why this threshold frequency was different for different metals

b why the number of photoelectrons depended on the intensity of the light incident on the metal surface but not on its frequency.

Einstein was able to show that a complete explanation of the photoelectric emission of electrons is possible if light is assumed to be quantised. If we consider a free electron, close to the inside surface of a piece of metal, the electron will require a certain amount of energy in order to escape completely from the metal. When photons of light strike the surface of the metal, one may strike the electron. As a result of this collision the photon's energy may be transferred to the electron, and it may now have sufficient energy to escape from the metal.

The emission of photoelectrons from a metal surface and their subsequent kinetic energy both depend on the frequency of the incident light. An electron absorbs energy from an incident photon in order to escape from the metal. Any energy remaining after the electron has done the work necessary to escape from the metal remains as the electron's kinetic energy.

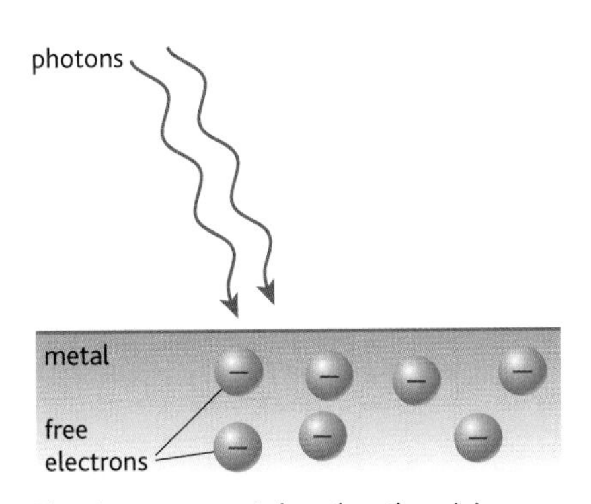

The photon energy is less than the minimum energy required to eject an electron from the metal – no photoelectrons are produced.

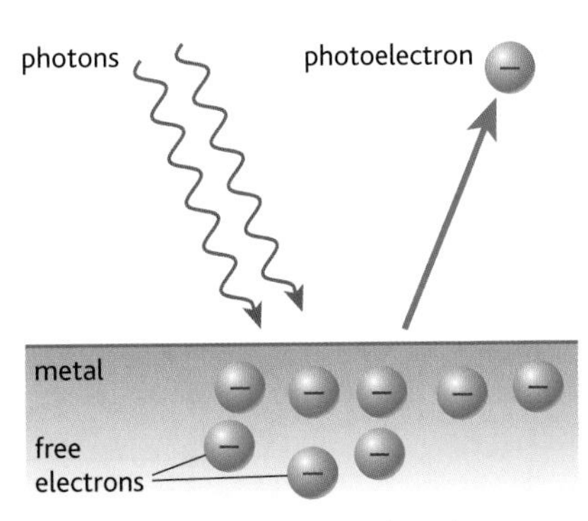

The photon energy is greater than the minimum energy required to eject an electron from the metal. Photoelectrons are produced which have a range of kinetic energies up to a maximum value.

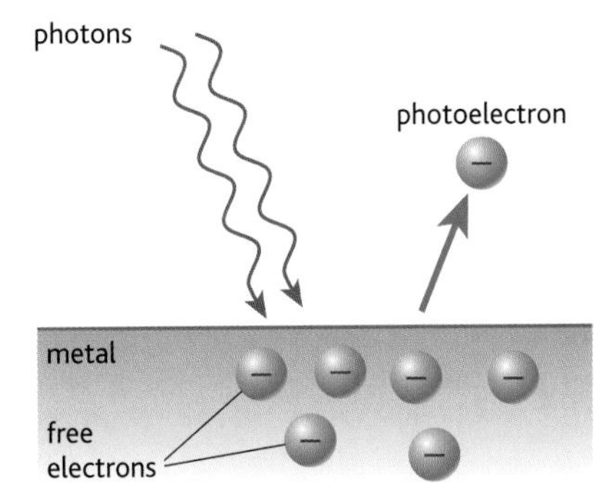

The photon energy is just large enough to cause emission. Photoelectrons with zero kinetic energy are produced.

fig. 5.1.6 A metal surface being bombarded by photons with three different energies.

Note that absorption of the energy required to escape from the metal surface must happen in a single interaction with a photon – no multiple collisions to acquire this energy are allowed.

Einstein's photoelectric equation expresses these ideas mathematically:

$$hf = \phi + \tfrac{1}{2}mv^2_{max}$$

The photon energy is given by hf as we have already seen, and $\tfrac{1}{2}mv^2_{max}$ is the maximum kinetic energy of the photoelectron. The quantity ϕ is the **work function** of the metal surface, which is the energy required to completely remove an electron from the metal. Fig. 5.1.7 shows how the maximum kinetic energy of electrons varies with the frequency of incident light for caesium.

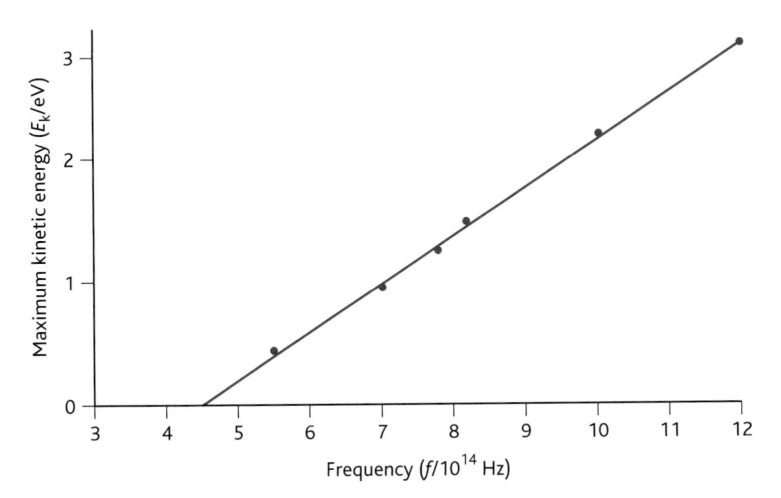

fig. 5.1.7 This graph shows the maximum kinetic energy for photoelectrons emitted from caesium by light of different frequencies. The energy is given in units of electronvolts (eV).

Energy and atoms – the electronvolt

When considering energy and subatomic particles it is convenient to use a unit of energy called the **electronvolt** (eV). 1 eV is the energy transferred when an electron travels through a potential difference of 1 volt. Since the size of the charge on the electron is 1.6×10^{-19} C, and energy transferred = charge × pd (see chapter 4.1):

$$1\,\text{eV} = 1.6 \times 10^{-19}\,\text{J}$$

Worked examples

a Use the graph in fig. 5.1.7 to obtain an estimate for the work function of caesium.

b Light of wavelength 300 nm is incident on a caesium surface. Calculate the maximum kinetic energy of the photoelectrons emitted from the surface. Express your answers to an appropriate degree of accuracy. ($c = 3.0 \times 10^8\,\text{m s}^{-1}$, $h = 6.63 \times 10^{-34}\,\text{J s}$.)

a An electron emerging from the metal with zero kinetic energy will have absorbed just enough energy to have escaped from the metal, so:

$hf = \phi + 0$

The work function can therefore be obtained from the graph in fig. 5.1.7 as the intercept on the x-axis, where $\frac{1}{2}mv^2_{max} = 0$.

intercept on x-axis = $4.5 \times 10^{14}\,\text{Hz}$

So:

$$\phi = 6.63 \times 10^{-34} \times 4.5 \times 10^{14}$$
$$= 3.0 \times 10^{-19}\,\text{J}$$

This answer is given to two significant figures, since it is impossible to read the graph with any greater accuracy.

b For light of incident wavelength 300 nm:

$$hf = \frac{hc}{\lambda} = \phi + \frac{1}{2}mv^2_{max}$$

Rearranging and substituting:

$$\frac{1}{2}mv^2_{max} = \frac{6.63 \times 10^{-34} \times 3.0 \times 10^8}{3.0 \times 10^{-7}} - 3.0 \times 10^{-19}$$
$$= 3.7 \times 10^{-19}\,\text{J}$$

(again to two significant figures)

The work function for caesium as calculated from the graph is 3.0×10^{-19} J, and the maximum kinetic energy of the photoelectrons is 3.7×10^{-19} J.

Note: It is common to express both work function and the kinetic energy of photoelectrons in electronvolts. These answers may be converted to electronvolts by dividing them by $1.6 \times 10^{-19}\,\text{J eV}^{-1}$, which gives:

$\phi = 1.9\,\text{eV}$ $\quad \frac{1}{2}mv^2_{max} = 2.3\,\text{eV}$

Investigating the photoelectric effect

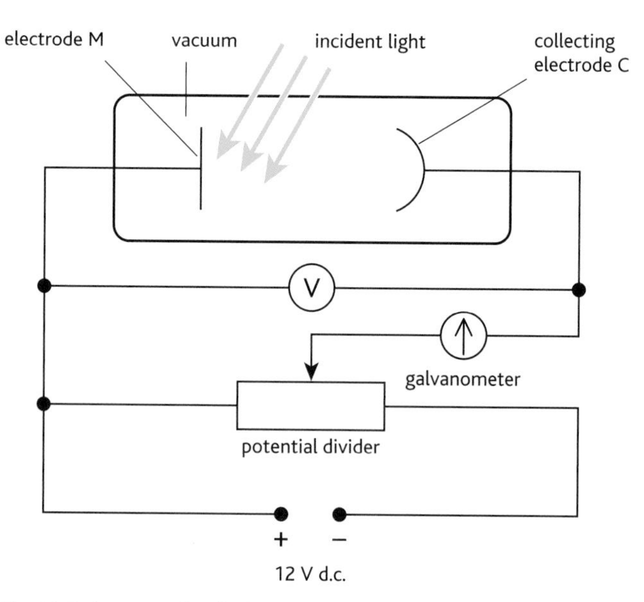

fig. 5.1.8 Apparatus for the investigation of the photoelectric effect. This apparatus also enables a value for Planck's constant to be obtained.

The investigation of the photoelectric effect may be carried out using the apparatus shown in fig. 5.1.8. Light of a known frequency is shone onto a metal electrode M. Photoelectrons from M travel towards a collecting electrode C and then flow round the external circuit through the galvanometer.

The maximum kinetic energy of the ejected electrons can be found by applying a potential difference across C and M, making C negative with respect to M. As this potential difference is increased (by means of the potential divider), the number of electrons reaching C from M will decrease as fewer and fewer electrons have sufficient energy to overcome the potential barrier. Eventually the potential difference will be such that even the most energetic electrons will just fail to reach C. This potential is called the **stopping potential**, V_{stop}. The measured value of this stopping potential gives the maximum kinetic energy of the electrons:

$$\frac{1}{2}mv^2_{max} = eV_{stop}$$

Using apparatus like this, the frequency of the incident light may be varied (a spectrometer may be used as a source of incident light, or alternatively filters admitting a narrow range of frequencies may be used) and the maximum kinetic energy of the electrons found from the stopping potential. This is the method used to obtain the graph shown in fig. 5.1.7.

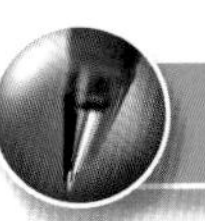

Worked examples

A metal surface with a work function of 2.86 eV is illuminated with light of wavelength 400 nm. What will be the measured stopping potential for the photoelectrons?

($h = 6.63 \times 10^{-34}$ J s, $c = 3.0 \times 10^{8}$ m s^{-1}, $e = -1.6 \times 10^{-19}$ C.)

The work function ϕ in joules is given by:

$$\phi = 2.86 \times 1.6 \times 10^{-19}$$
$$= 4.58 \times 10^{-19}\,\text{J}$$
$$hf = \frac{hc}{\lambda} = \phi + \tfrac{1}{2}mv^2_{max}$$
$$= \phi + eV_{stop}$$

Rearranging and substituting:

$$V_{stop} = \frac{(6.63 \times 10^{-34} \times 3.0 \times 10^{8}}{4.0 \times 10^{-7}} - 4.58 \times 10^{-19})$$
$$\times \frac{1}{1.6 \times 10^{-19}} = 0.25\,\text{V}$$

Questions

1 Explain why zinc will emit photoelectrons if it is illuminated with a dim source of ultraviolet light, but will not emit any photoelectrons when illuminated with a very bright red light.

2 a State Einstein's photoelectric equation and explain all the terms.

b Light with a wavelength of 0.5 μm causes photoelectrons to be emitted from a piece of potassium (work function $\phi = 2.30$ eV). What is the maximum kinetic energy of these electrons?

c What is the maximum speed of the photoelectrons in part **b**?

d What is the threshold frequency for emission of photoelectrons from potassium?

3 Describe a photoelectric effect experiment to determine Planck's constant. Include details of measurements to be made and how these results can be analysed to find h. Also describe how any further quantities can be obtained from analysis of the results.

4 Give an example of how the photomultiplier as used in a camera developed for low light levels could be used in society to inform decision making.

HSW Remote sensing

The photoelectric effect has practical applications in a number of areas concerned with the detection of light. In the **photomultiplier tube** a single incident photon produces a pulse of current through an external circuit as a result of a series of electron avalanches (**fig. 5.1.9**). A high gain tube may produce as many as 10^9 electrons from a single photon.

This type of tube can be used in an image intensifying camera to obtain pictures in extremely low light levels (**fig. 5.1.10**), or in a scintillation counter used to detect ionising events. In the case of the scintillation counter the photon of light comes from a crystal of sodium iodide which emits a weak flash of light when ionising radiation passes through it.

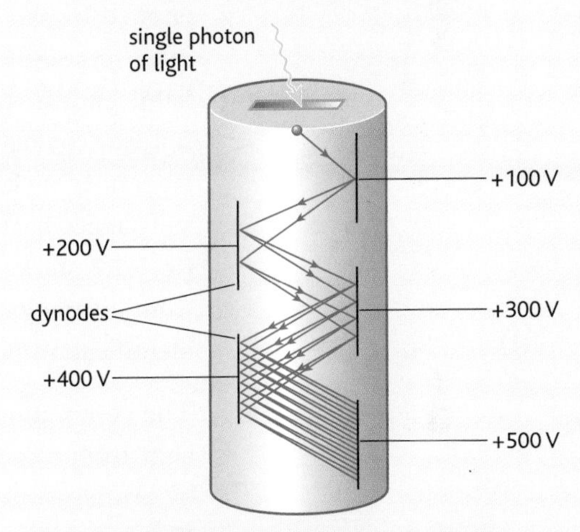

fig. 5.1.9 A single photon hitting the first electrode causes the emission of a photoelectron. This is accelerated towards the first of a series of secondary electrodes called dynodes, where further electrons are released. In this diagram each electron impact on a dynode has been assumed to cause the release of two secondary electrons.

fig. 5.1.10 Image taken using an image-intensifying camera.

5.2 Spectra and energy levels in atoms

Types of spectra

In chapter 3.3 we saw how different wavelengths of electromagnetic radiation form a complete spectrum. A **spectrometer** is a device that can separate the wavelengths in a beam of radiation, to show those wavelengths that are present. In experiments with the spectrometer we can see that two different types of spectra appear – one consisting of lines, the other of a continuous range of colours. These two types of spectra are named from their appearance – **line spectra** and **continuous spectra** respectively, and result from the way in which the electrons in matter are able to radiate energy. Because these spectra result from the emission of energy they are called **emission spectra**.

Seen through a spectrometer, line emission spectra consist of a set of coloured lines against a dark background, each line being a particular wavelength of light emitted by the source. These lines also extend beyond the visible spectrum in many cases, to include the infrared (wavelengths longer than those of visible light) and the ultraviolet (wavelengths shorter than those of visible light), as in the full electromagnetic spectrum. Convenient sources of line spectra to use in the laboratory include gases contained at low pressure in electric discharge tubes and substances heated in a blue Bunsen flame. Fig. 5.2.1 shows a gas discharge tube and fig. 5.2.2 shows the spectra obtained from different elements. The line spectrum from a sample of an element is characteristic of that element, and can be used for identification purposes.

The continuous range of colours seen in continuous spectra can be observed using light from a tungsten filament lamp. This type of spectrum was first produced by artificial means (as opposed to by raindrops falling through air) by Sir Isaac Newton, using a prism. Unlike line spectra, continuous spectra cannot be used to identify their source, although the wavelength of maximum intensity of the spectrum is linked to the temperature of the source.

The final example of spectra in fig. 5.2.2 is due to the absorption of light by a sample, and is therefore an example of an **absorption spectrum**. In this case, light from a tungsten filament source has been passed through sodium atoms in the vapour phase. In place of the two strong lines of yellow light there appear two black lines that are due to the absorption of light by sodium atoms at exactly the same wavelength.

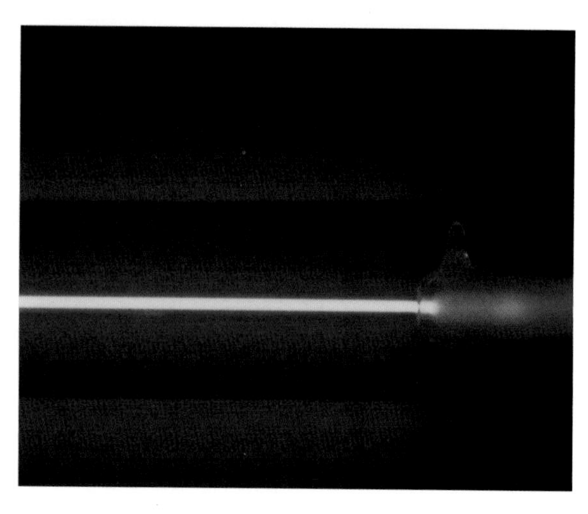

fig 5.2.1 A gas discharge tube.

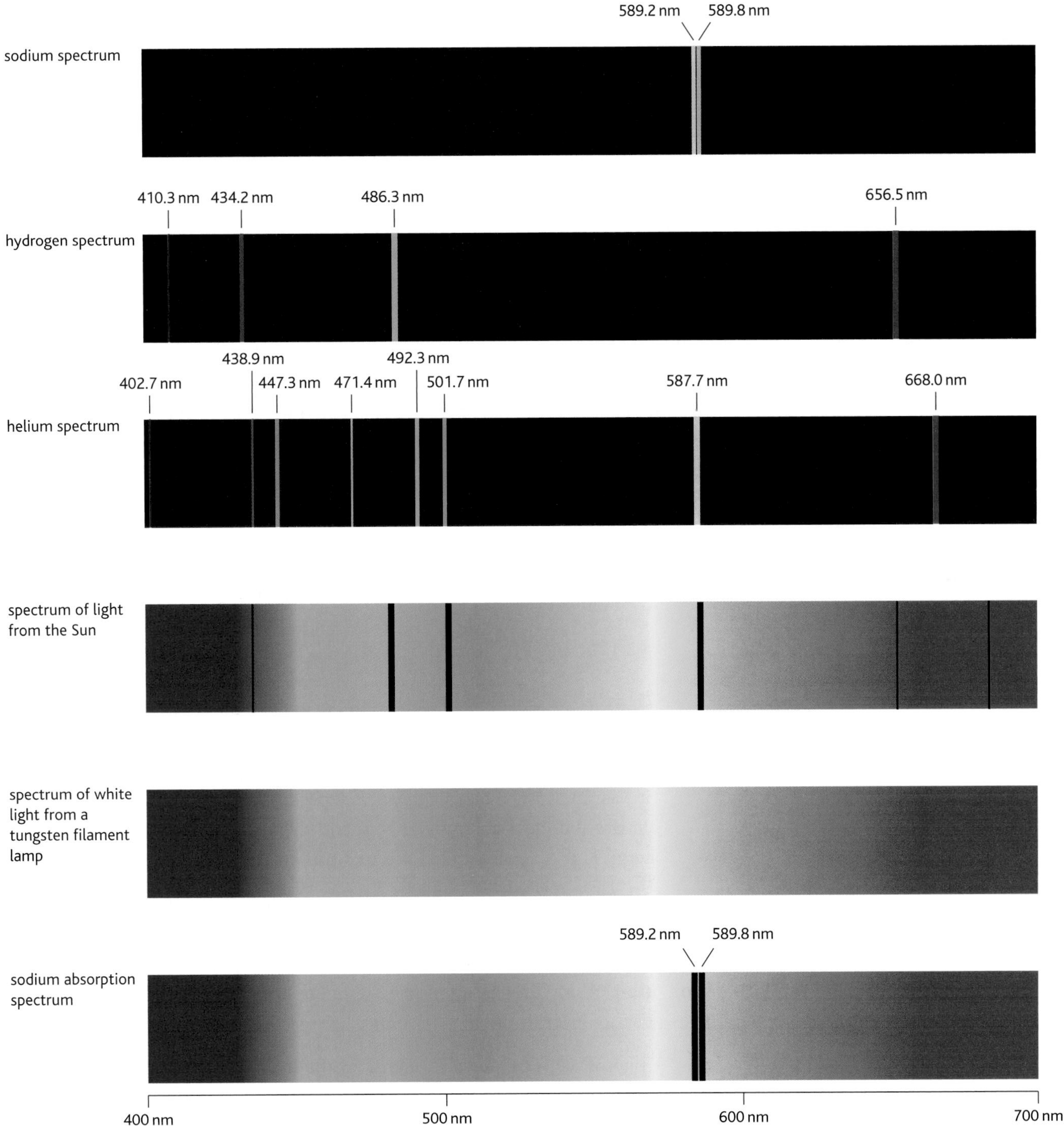

fig. 5.2.2 Spectra from a range of different sources. The dark lines in the spectrum from the Sun are called Fraunhofer lines.

The Solar spectrum

The presence of dark lines in the spectrum of sunlight was first recorded by William Hyde Woolaston in 1802, but it was Joseph von Fraunhofer who realised their significance, recording the lines in his map of the solar spectrum (containing 324 lines) published in 1815.

The spectrum of light from the Sun in **fig. 5.2.2** is a simplified one with only the main Fraunhofer (dark) lines, but there is clearly strong absorption at wavelengths corresponding to the lines in the emission spectra of hydrogen and helium. This shows the presence of these elements in the mantle of hot gas around the Sun. It was the careful analysis of these lines that led to the discovery of helium by Joseph Lockyer some 50 years later. (The name helium comes from the Greek name for the Sun, *helios*).

Atomic electron energies

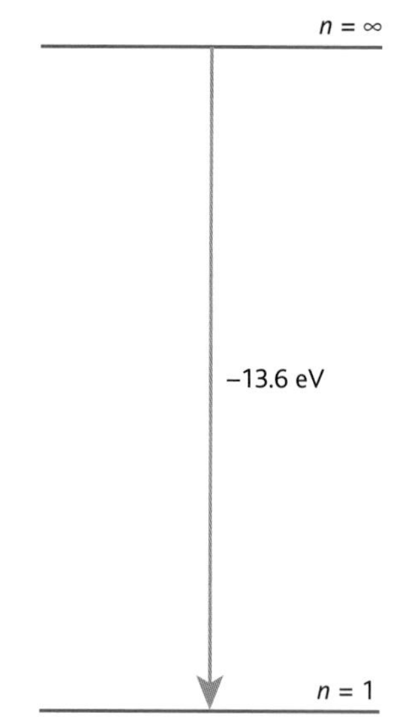

fig. 5.2.3 In going from $n = \infty$ to $n = 1$ the atom loses 13.6 eV of energy and releases a photon with 13.6 eV of energy.

Energy levels and the production of spectra

Why do the spectra of elements only show emission and absorption of light at particular wavelengths? From the photoelectric effect and Einstein's photon theory, we know that a photon with a particular wavelength represents a fixed amount of energy. So each line in an emission spectrum corresponds to the atom losing a fixed amount of energy, given out as a photon of light.

The element hydrogen consists of one proton and one electron. The shortest wavelength in the hydrogen spectrum is 9.17×10^{-8} m, and this light also has the highest frequency, and so represents the most energetic photon that can be released. This wavelength corresponds to an energy change ΔE given by:

$$E = hf = \frac{hc}{\lambda}$$

$$E = \frac{6.63 \times 10^{-34} \times 3 \times 10^{8}}{9.17 \times 10^{-8}}$$

$$= 2.17 \times 10^{-18}\,\text{J}$$

$$= 13.6\,\text{eV}$$

As this is the largest energy change in the spectrum, it must correspond to the electron moving from a state where the atom has the most energy to one where it has the least. This is shown in fig. 5.2.3, where the state in which the atom has the most energy is $n = \infty$ and the state in which the atom has least energy is $n = 1$, also called the **ground state**.

The same amount of energy, 13.6 eV, is absorbed by a hydrogen atom when it is ionised from its ground state, which is when the electron is given enough energy to escape from the attraction of the nucleus altogether. Thus, **ionisation** of a hydrogen atom in its ground state is a transition from the $n = 1$ energy level to the $n = \infty$ energy level.

So the photon with the highest energy in the hydrogen spectrum, and thus the shortest wavelength, corresponds to a free electron moving to the lowest energy level, with an energy change of −13.6 eV (energy is given out). But the spectrum has many lines of longer wavelength, which correspond to electron transitions with smaller energy changes. These correspond to transitions between intermediate energy levels, as shown in fig. 5.2.3. The spectral line with the longest wavelength (and so the least energy) from a transition involving the ground state is due to a transition between the levels $n = 1$ and $n = 2$.

Looking at the energy-level diagram in this way, you can see that further lines will be produced by other transitions between energy levels, such as a drop from level 4 to level 2. The wavelength produced will still correspond to the energy difference between the two levels, following Einstein's equation for photon energy.

When the atom absorbs energy, and the electron moves from a lower energy level to a higher one (excitation), it can only absorb energy that corresponds exactly to the energy difference between two energy levels. So absorption also only occurs for fixed wavelengths of light.

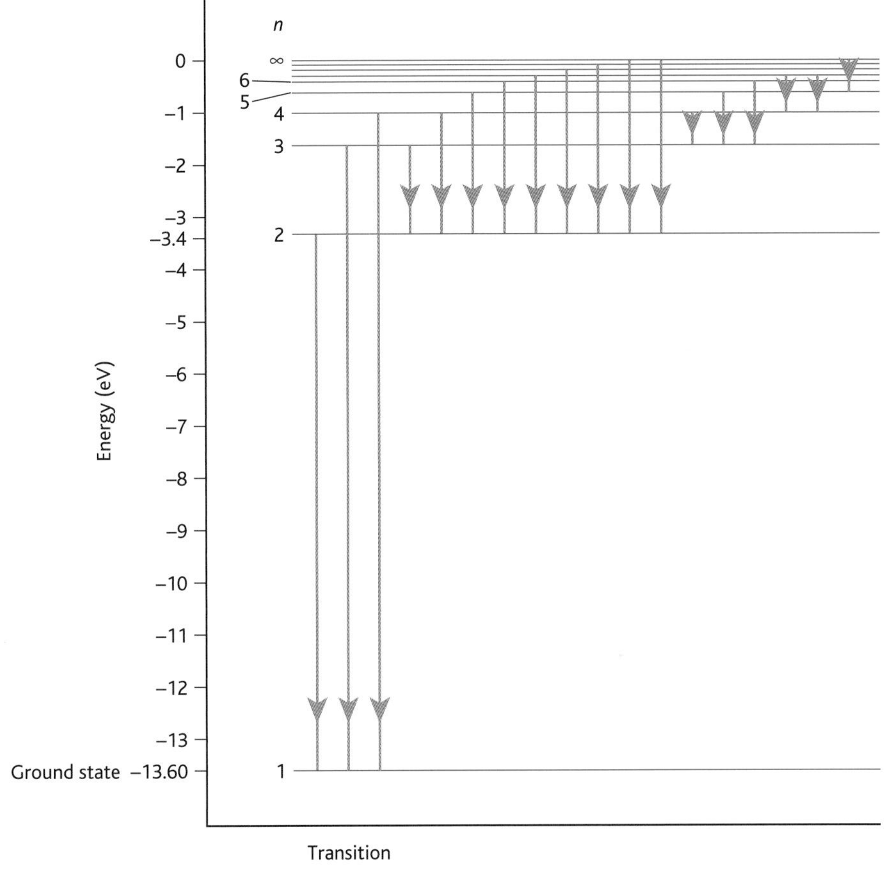

fig. 5.2.4 **Energy levels in the hydrogen atom and the production of spectral lines.**

This approach can explain the most important aspects of atomic spectra:

1 The production of lines is due to the existence of distinct energy levels within the atom.

2 A unique line is produced for each transition between any two electron energy levels within the atom, and the frequency/wavelength of that line corresponds to the energy difference between those two levels.

3 Absorption spectra (seen for example in the spectrum of light from the Sun) are due to the absorption of light and the transition of an electron within the atom from a lower energy level to a higher one.

Because the precise energy-level structure is different for each element, each element's line spectrum is a unique fingerprint that identifies that element. This principle is used to determine the chemical composition of stars.

HSW Franck and Hertz's experiment

Evidence that absorption and emission of radiation are concerned with the movement of electrons in atoms came from experiments carried out by Franck and Hertz in 1913.

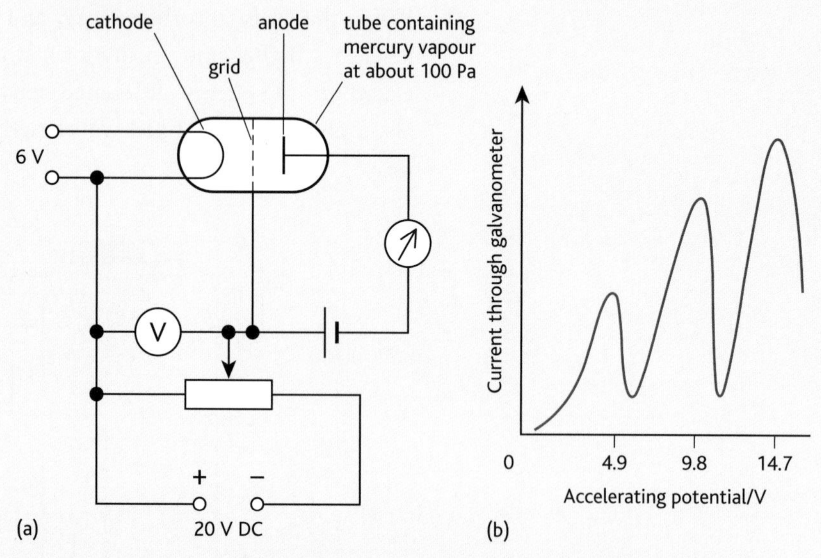

fig. 5.2.5 **Franck and Hertz's experiment used a tube containing low-pressure mercury vapour. Triode valves of this type used to be commonly found in radio sets until the advent of the transistor.**

In this experiment electrons from the cathode were accelerated towards the grid. These electrons then continued towards the anode, from which they could flow back to the power supply, via a galvanometer. A small stopping potential of about 2 V was applied between the grid and the anode – i.e. the anode has a potential about 2 V less than the grid.

Fig. 5.2.5b shows the results of the experiment. As the accelerating potential (the potential *V* between the cathode and the grid) was increased the current through the tube increased, reaching an initial maximum at 4.9 V and then decreasing sharply. Further increasing the accelerating potential caused the current to rise again, when it peaked again at 9.8 V and then fell. Another peak at 14.7 V was also seen. At the same time as these peaks of current occurred the mercury vapour glowed.

Franck and Hertz explained these observations in terms of the energy levels of electrons in the mercury atom. With an accelerating potential below 4.9 V, electrons collided elastically with mercury atoms in their path. However, electrons accelerated by a potential of 4.9 V underwent inelastic collisions with mercury atoms, having just enough energy to excite an electron in the mercury atom to a higher energy level and in the process lose all of their own energy. As the accelerating potential was increased above 4.9 V electrons had some energy remaining after an inelastic collision, so could overcome the stopping potential between the grid and anode. The subsequent peaks at 9.8 V and 14.7 V corresponded to electrons having sufficient energy to excite two and three mercury atoms, respectively.

A similar arrangement can be used to measure the ionisation energy of an atom.

Laser light and energy transitions

Laser light can be used to produce a spectacular display (fig. 5.2.7). Laser light is emitted when atoms undergo similar energy transitions at the same time. This is achieved by promoting a large number of atoms to an energy level above the ground state. As an electron in one of the excited atoms jumps down from its higher energy level it emits a photon. As this photon travels past another atom in an excited state it causes the electron in this atom to jump down to the lower level. The passage of light thus encourages or **stimulates** the emission of radiation from other atoms – producing the intense, coherent beam of light characteristic of the laser.

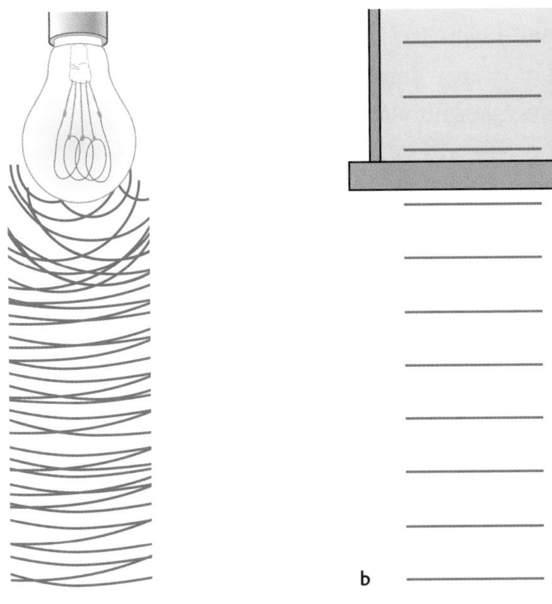

fig. 5.2.6 a In a source like a tungsten filament lamp the atoms emit light as they move from higher levels to lower levels quite independently. The result is a jumbled combination of waves. b In a laser source atoms make their jumps between energy levels together. The light that this produces consists of a coherent combination of waves.

fig. 5.2.7 Lasers can produce a spectacular display.

Questions

1 Distinguish between:

  a line spectra and continuous spectra

  b emission spectra and absorption spectra

  c excitation and ionisation.

2 Calculate the wavelength of a photon of light emitted from a hydrogen atom as a result of an electron transition from level $n = 2$ to the ground state (look at fig. 5.2.4 for energy data).

3 Why is the line spectrum from an atom useful to identify what element it is? Give an example of when you may have used this in a chemistry lesson.

4 What would happen if a photon with an energy of 10 eV were incident on an unexcited hydrogen atom? What about a 20 eV photon? How about two 10 eV photons?

Solar cells to light the world?

Fossil fuels are becoming increasingly expensive as the resources run out. Their use also damages the environment in numerous ways, in particular, causing climate change through emission of carbon dioxide into the atmosphere. Sunlight is a free source of energy. It is clean and will never run out. Why is it, then, that society's energy usage from the Sun remains minimal?

fig. 5.2.8 Could the Sun's energy be harnessed to replace use of fossil fuels?

Solar energy can be harnessed in a number of ways. Most familiar in the UK is the photovoltaic cell, which produces an electric current directly from light. This is commonly the power source for calculators. Other systems in which water is heated from the light energy, either to be used as hot water or to drive turbines and generate electricity, are also feasible.

fig. 5.2.9 Photovoltaic cells can supply electricity directly from sunlight.

Photovoltaic efficiency

Photovoltaic cells produce an electric current through the interaction of light photons with semiconductor materials. In certain materials, like silicon, a photon may excite an electron from an energy level where it is fixed to an atom up to one in which it can move through the material as a conduction electron. The exact chemical composition of the semiconductor material will determine the photon energy required to excite electrons into the conduction band. The chemical composition will also affect the cost. Cost, in comparison with the quantity of energy delivered, is critically important when comparing solar energy with alternatives. This means the efficiency of the photovoltaic cells will be crucial in determining whether people will use them in preference to fossil fuels.

Efficiency is defined as the ratio of energy output to energy input. If this is considered in a certain time period, then it is also the ratio of power output to power input:

$$\text{efficiency} = \frac{\text{useful energy (or power) output}}{\text{total energy (or power) input}} \times 100\%$$

Worked example

A new type of 'hybrid' photovoltaic cell has an area of $0.5\,\text{m}^2$ and produces 82 W of electricity. What is its efficiency?

The radiation from the Sun lands on Earth's surface at a flux of $1000\,\text{W}\,\text{m}^{-2}$. So half a square metre receives 500 W.

$$\text{efficiency} = \frac{82}{500} \times 100\%$$

$$= 16.4\%$$

Efficiency affects the electricity output of a photovoltaic cell, and is also closely connected with cost. Other cost factors, such as the extent of cabling required, can make solar power the only viable solution in some situations. An emergency telephone by a road in the middle of the Sahara Desert could not be viably powered from a power station 1000 kilometres away.

How does society choose its energy sources?

Science does not aim to answer ethical questions, but to provide accurate data, presented in a balanced, objective way so that individuals and society can make decisions based on the most up

to date and accurate facts known. Society's choices about changing energy use to solar will mostly be influenced by information provided by scientists. Important considerations include:

- cost – the market for solar photovoltaic cells is expanding and, as production increases, the price is falling
- global energy prices, especially for fossil fuels, are rising
- the rise in national legislation and international commitments designed to reduce carbon emissions
- a realisation, and acceptance, of the risks associated with global warming.

Case study: The CIS tower Manchester

The CIS tower in Manchester recently underwent an extensive and costly refurbishment. One aspect of this was a £5.5 million refit of the exterior tiling on the service tower. It was all replaced with photovoltaic cells. Externally, it seems similar to most skyscrapers, with the reflectiveness of the cells adding something of a space-age look.

fig. 5.2.10 Using photovoltaic cells as a building material reduces their capital cost as the costs replace those of the alternative materials.

The solar cells link into the National Grid and provide enough electricity to power about 300 homes. At current electricity prices this means that the £5.5m will be recouped in about 30 years. This is not an unreasonable timescale given that the original tiling had been in place since 1962 (and that some of the cost was borne by government grants). However, most building projects considering the use of photovoltaic building materials – they can replace roof tiles, or tinted-glass windows – would want a quicker return than this. If the price drops further, more people may be able to afford to follow their conscience.

HSW Photovoltaic technology for the future

In order to increase society's use of solar cells, they need to be developed to be more efficient, and cheaper. Also, there is some concern that they may have a significant impact on the environment that has, as yet, been largely ignored. The mining and refining of the semiconductor materials used to manufacture photovoltaic cells is an energy-intensive process, meaning that their production will put a strain on resources. This 'energy cost' also needs to be reduced.

The latest developments in photovoltaic technology include research in the following areas:

- Using electrolysis, at a lower temperature than previously possible, it has been discovered that solid silica can be converted into pure silicon. Very pure silicon is essential in the manufacture of some photovoltaic cells and this could significantly reduce production costs.
- The use of silicon may be largely removed if research into mass production of thin-film technologies continues to show dramatic improvements. These require 99% less silicon than the wafer-based design of solar cells used to date.
- Also cheaper would be to replace silicon and use inexpensive plastics. This may become possible with the development of conductive polymers.
- Photovoltaic cells function over a very limited range of wavelengths (only 23% of electromagnetic waves reaching the Earth's surface are at an appropriate wavelength). Chemicals that can absorb a much wider range of wavelengths and then re-emit them at useful wavelengths for photovoltaic action will improve efficiency significantly.

Questions

1. Why are solar cells more environmentally friendly than generating electricity with fossil fuels?
2. What factors will affect whether people switch to using solar cells rather than other energy sources?
3. Calculate the power output of a solar cell array that is made from 20 photovoltaic tiles. Each tile is 20 cm square and has an efficiency of 12%.

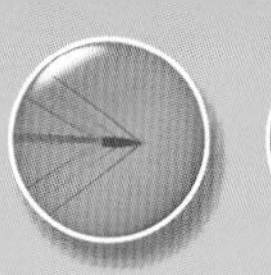
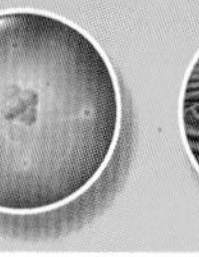

Index

Pearson Education Limited
Edinburgh Gate
Harlow
Essex CM20 2JE

The right of Miles Hudson and Patrick Fullick to be identified as the authors of this work has been asserted by him/them in accordance with the Copyright, Designs and Patents Act of 1988.

First published 2008

Fourth impression 2009

ISBN 978-1-4058-9638-2

Design 320 Design Ltd.
Illustration Oxford Designers and Illustrators
Picture research Kay Altwegg
Index John Holmes
Printed and bound in China GCC/04

The publisher's policy is to use paper manufactured from sustainable forests.

Acknowledgements
The publishers are grateful to Richard Laird for his collaboration in reviewing this book.

Photo acknowledgements
The publisher would like to thank the following for their kind permission to reproduce their photographs:

(Key: b-bottom; c-centre; l-left; r-right; t-top)

13 Getty Images: AFP. 24 Science Photo Library Ltd: DR. JEREMY BURGESS. 25 Alamy Images: Mark Scheuern. 28 Science Photo Library Ltd: NASA. 29 Alamy Images: sciencephotos (t). Science Photo Library Ltd: ANDREW LAMBERT PHOTOGRAPHY (b). 39 Science Photo Library Ltd: MEHAU KULYK. 41 Alamy Images: Digital Vision (b); JUPITERIMAGES/ Thinkstock (t). Science Photo Library Ltd: SHEILA TERRY. 44 Alamy Images: Greg Balfour Evans (b); Robert Towell (t). 46 PA Photos: AP Photo/Gazpar Latorre (l); AP Photo/M. Lakshman (r). 52 Getty Images: Gunnar Smoliansky. 54 Alamy Images: MARTIN DALTON. 55 Trevor Clifford: Pearson Education. 56 Alamy Images: blickwinkel. 57 Trevor Clifford: Pearson Education (b). Robert Harding World Imagery: James Emmerson (cl); Amanda Hall (cr). Science Photo Library Ltd: TAKESHI TAKAHARA (tl). 58 Getty Images: Jonathan Wood. 60 iStockphoto: (l). Science Photo Library Ltd: (r). 65 Alamy Images: mediacolor's (b). iStockphoto: (t). 66 Trevor Clifford: Pearson Education. 68 Alamy Images: sciencephotos. 70 Alamy Images: imagebroker (t). GeoScience Features Picture Library: (b). 72 Pearson Education Ltd: (l) (r). 74 Science Photo Library Ltd: Maximilian Stock Ltd. 79 Alamy Images: sciencephotos (r) (l). 84 Alamy Images: moodboard (t). Science Photo Library Ltd: ANDREW LAMBERT PHOTOGRAPHY (b). 89 iStockphoto: (r) (l). 90 Alamy Images: sciencephotos (l) (r). 92 Alamy Images: sciencephotos (r) (l). 93 Alamy Images: sciencephotos (b) (t). 95 Trevor Clifford: Pearson Education. 96 Alamy Images: sciencephotos (tc); Stock Connection Distribution (t). Science Photo Library Ltd: ANDREW LAMBERT PHOTOGRAPHY (b) (bc). 97 Alamy Images: sciencephotos (b). Science Photo Library Ltd: Science Source (t). 99 Alamy Images: JUPITERIMAGES/ Comstock Images. 101 Science Photo Library Ltd: DR ARTHUR LESK. 104 iStockphoto: (t). Pearson Education Ltd: (bc) (c) (tc) (b). 106 Alamy Images: fstop2 (t). Science Photo Library Ltd: GUSTOIMAGES (br); JERRY MASON (bl). 107 iStockphoto: (b) (t). 109 Science Photo Library Ltd: JIM AMOS. 111 Alamy Images: Chad Ehlers (l). Science Photo Library Ltd: Zephyr (r). 119 Mary Evans Picture Library: (tr). Science Photo Library Ltd: (bl) (tl) (2/cr); J-L CHARMET (cr); LIBRARY OF CONGRESS (br); SHEILA TERRY (br/centre) (bl/centre) (cl) (2/cl). Wellcome Images: (tc). 126 Camera Press Ltd: Photograph by Erma. 130 Ordnance Survey. 135 Trevor Clifford: Pearson Education (l) (r). 136 Alamy Images: Transtock Inc.. 140 Trevor Clifford: Pearson Education. 141 Alamy Images: Chris Howes/Wild Places Photography. 149 Mary Evans Picture Library: (tr) (bl) (tc) (tl) (2/bc). Science Photo Library Ltd: (br); BILL SANDERSON (1/bc). 152 Mary Evans Picture Library: Weimar Archive. 155 Science Photo Library Ltd. 156 Science Photo Library Ltd: RICH TREPTOW. 161 Alamy Images: imagebroker. 162 Alamy Images: JIM WEST (t); Maximilian Weinzierl (b). 163 Alamy Images: Andrew Jankunas

All other images © Pearson Education

Picture Research by: Kay Altwegg

Every effort has been made to trace the copyright holders and we apologise in advance for any unintentional omissions. We would be pleased to insert the appropriate acknowledgement in any subsequent edition of this publication.

Single User Licence Agreement:
Edexcel AS Physics Student Book with FREE ActiveBook CD-ROM

Warning:
This is a legally binding agreement between You (the user or purchasing institution) and Pearson Education Limited of Edinburgh Gate, Harlow, Essex, CM20 2JE, United Kingdom ('PEL').

By retaining this Licence, any software media or accompanying written materials or carrying out any of the permitted activities You are agreeing to be bound by the terms and conditions of this Licence. If You do not agree to the terms and conditions of this Licence, do not continue to use the Edexcel AS Physics Student Book with FREE ActiveBook CD-ROM and promptly return the entire publication (this Licence and all software, written materials, packaging and any other component received with it) with Your sales receipt to Your supplier for a full refund.

Intellectual Property Rights:
This Edexcel AS Physics Student Book with FREE ActiveBook CD-ROM consists of copyright software and data. All intellectual property rights, including the copyright is owned by PEL or its licensors and shall remain vested in them at all times. You only own the disk on which the software is supplied. If You do not continue to do only what You are allowed to do as contained in this Licence you will be in breach of the Licence and PEL shall have the right to terminate this Licence by written notice and take action to recover from you any damages suffered by PEL as a result of your breach.

The PEL name, PEL logo, Edexcel name, Edexcel logo and all other trademarks appearing on the software and Edexcel AS Physics Student Book with FREE ActiveBook CD-ROM are trademarks of PEL. You shall not utilise any such trademarks for any purpose whatsoever other than as they appear on the software and Edexcel AS Physics Student Book with FREE ActiveBook CD-ROM.

Yes, You can:
1. use this Edexcel AS Physics Student Book with FREE ActiveBook CD-ROM on Your own personal computer as a single individual user. You may make a copy of the Edexcel AS Physics Student Book with FREE ActiveBook CD-ROM in machine readable form for backup purposes only. The backup copy must include all copyright information contained in the original.

No, You cannot:
1. copy this Edexcel AS Physics Student Book with FREE ActiveBook CD-ROM (other than making one copy for back-up purposes as set out in the Yes, You can table above);

2. alter, disassemble, or modify this Edexcel AS Physics Student Book with FREE ActiveBook CD- ROM, or in any way reverse engineer, decompile or create a derivative product from the contents of the database or any software included in it:

3. include any materials or software data from the Edexcel AS Physics Student Book with FREE ActiveBook CD-ROM in any other product or software materials;

4. rent, hire, lend, sub-licence or sell the Edexcel AS Physics Student Book with FREE ActiveBook CD-ROM;

5. copy any part of the documentation except where specifically indicated otherwise;

6. use the software in any way not specified above without the prior written consent of PEL;

7. subject the software, Edexcel AS Physics Student Book with FREE ActiveBook CD-ROM or any PEL content to any derogatory treatment or use them in such a way that would bring PEL into disrepute or cause PEL to incur liability to any third party.

Grant of Licence:
PEL grants You, provided You only do what is allowed under the 'Yes, You can' table above, and do nothing under the 'No, You cannot' table above, a non-exclusive, non-transferable Licence to use this Edexcel AS Physics Student Book with FREE ActiveBook CD-ROM.

The terms and conditions of this Licence become operative when using this Edexcel AS Physics Student Book with FREE ActiveBook CD-ROM.

Limited Warranty:
PEL warrants that the disk or CD-ROM on which the software is supplied is free from defects in material and workmanship in normal use for ninety (90) days from the date You receive it. This warranty is limited to You and is not transferable.

This limited warranty is void if any damage has resulted from accident, abuse, misapplication, service or modification by someone other than PEL. In no event shall PEL be liable for any damages whatsoever arising out of installation of the software, even if advised of the possibility of such damages. PEL will not be liable for any loss or damage of any nature suffered by any party as a result of reliance upon or reproduction of any errors in the content of the publication.

PEL does not warrant that the functions of the software meet Your requirements or that the media is compatible with any computer system on which it is used or that the operation of the software will be unlimited or error free. You assume responsibility for selecting the software to achieve Your intended
results and for the installation of, the use of and the results obtained from the software.

PEL shall not be liable for any loss or damage of any kind (except for personal injury or death) arising from the use of this Edexcel AS Physics Student Book with FREE ActiveBook CD-ROM or from errors, deficiencies or faults therein, whether such loss or damage is caused by negligence or otherwise.

The entire liability of PEL and your only remedy shall be replacement free of charge of the components that do not meet this warranty.
No information or advice (oral, written or otherwise) given by PEL or PEL's agents shall create a warranty or in any way increase the scope of this warranty.

To the extent the law permits, PEL disclaims all other warranties, either express or implied, including by way of example and not limitation, warranties of merchantability and fitness for a particular purpose in respect of this Edexcel AS Physics Student Book with FREE ActiveBook CD-ROM.

Termination:
This Licence shall automatically terminate without notice from PEL if You fail to comply with any of its provisions or the purchasing institution becomes insolvent or subject to receivership, liquidation or similar external administration. PEL may also terminate this Licence by notice in writing. Upon termination for whatever reason You agree to destroy the Edexcel AS Physics Student Book with FREE ActiveBook CD-ROM and any back-up copies and delete any part of the Edexcel AS Physics Student Book with FREE ActiveBook CD-ROM stored on your computer.

Governing Law:
This Licence will be governed by and construed in accordance with English law.